Bog-Trotting for Orchids

The Pink Moccasin-Flower. (*Cypripedium acaule.*)

Bog-Trotting for Orchids

Grace Greylock Niles

"I enter a swamp as a sacred place."—Thoreau

Coachwhip Publications
Landisville, Pennsylvania

Bog-Trotting for Orchids, by Grace Greylock Niles
Illustrated from photographs by Katherine Lewers and the author.
First published 1904. Reprinted with minor editing.

CoachwhipBooks.com

Front cover image: Philip Puleo
Back cover image: Fred Clark

ISBN 1-930585-53-5
ISBN-13 978-1-930585-53-9

Contents

Preface 7

First Season

1 Off to the Hills of Berkshire and Bennington 11
2 Ball Brook and the Bogs of Etchowog 20
3 The Haunts of the Ram's-Head Moccasin-Flowers 44
4 The Stolen Moccasins 48
5 The Queen of the Indian Moccasin-Flowers 57
6 Hail-Storms at Etchowog 70
7 Sweet Pogonias and Limodorums 79
8 A Colony of Ram's-Heads in Witch Hollow 90
9 Over the Huckleberry Plains 106

Second Season

10 Westville Swamps and Mount Carmel, Connecticut 113
11 May Showers and White Moccasin- Flowers 122
12 Saucy Jays and Polypores 130

Third Season

13 The Swamps and Hills of Mosholu and Lowerre 135
14 The Swamp of Oracles—Hoosac Valley 146
15 White Oaks and Gregor Rocks 158
16 Alpine Blossoms of the Dome 178
17 The Cascade and Bellows-Pipe, Notch Valley 185
18 The Natural Bridge of Mayunsook Valley 195
19 Orange Mountains, and Salt Meadows, New Jersey 200

Appendix: New England Orchids 209

One true-born blossom, native to our skies,
We dare not claim as kin,
Nor frankly seek, for all that in it lies,
The Indian's moccasin.

Elaine Goodale

Preface

During many seasons spent in the Hoosac Valley, it has been a source of great pleasure to me to trace mountain streams through moss-grown ravines to their beginnings, and to explore the almost inaccessible recesses of the sphagnous boglands. I have found it a delight to study the orchids, ferns, and various flowers sheltered in their homes, far removed from the roadside. I seldom follow any well-worn forest paths, for I have observed that the rarer plants do not dwell where the foot of man or the grazing herds have wandered. So it happens that the walks described in these pages lead mostly across lots, over hills and mountains, and through swamps.

The Hoosac Valley lies in the heart of the irregular Taconic Mountains, and extends over the southwestern part of Bennington County, Vermont, and the northwestern part of Berkshire County, Massachusetts. This region has a soil peculiarly adapted to the origin and growth of orchids. Here along the numerous streams and in the little vales are many unfathomable peat and marl beds which are veritable orchid gardens. The valley seems to be the common ground where rare plants from the North and South, as well as the migrating species from the East and West, meet and overlap each other.

Many people are accustomed to think of the orchid as a tropical flower which grows in our country only in cultivation and under highly artificial conditions. It is, however, true that many of the most attractive species of this beautiful group are endemic to most parts of the United States. There are to-day, according to conservative reports, from twenty-seven to thirty genera and from one hundred and fifty to one hundred and sixty species of native orchids found in North America, north of Mexico. Most of these are terrestrial or earth-loving. There are eleven epiphytes, all of which are found only in the Southern States. The range of the North American orchids extends wherever sunshine and moisture prevail, nearly as far north as the Arctic Circle. Four Cypripediums grow between latitudes 54° and 64°, and from fifteen to eighteen species of the Orchid Family are natives of Alaska.

The North Atlantic region, covering northeastern United States and Canada, produces seventy-one species of Orchidaceae; of these from forty-eight to fifty-six are reported for New England, and from forty to forty-two are found in the Hoosac Valley. Of the seventy-one North Atlantic orchids only fifteen or sixteen have not been found within Vermont. The most widely-known genus—*Cypripedium*, or Mocassin-Flower—is represented by thirteen species on the North American continent. This includes the single Mexican species. Six of this number have been collected in Connecticut, and five grow in the Hoosac Valley.

The excursions which I have recorded in this book were made particularly in search of orchids; but I have collected and observed all other flowers of interest which grow in the region which I have traversed, for the purpose of showing the natural environments of orchids, and introducing their near neighbors of swamp, forest, and rocky pasture-land.

G. G. N.
Williamstown, Massachusetts.

First Season

The Large Yellow Moccasin-Flower. (*Cypripedium hirsutum.*)

This common *Cypripedium* is closely allied with the Small Yellow Fragrant species—*Cypripedium parviflorum*—with which it grows in close comradeship, often intergrading. It is also nearly related with the European Yellow Cypripedium (*Cypripedium calceolus*), the first *Cypripedium* described by Linnæus in 1740-1753.

1
Off to the Hills of Berkshire and Bennington

> It is not the walking merely, it is keeping yourself in tune for a walk, in the spiritual and bodily condition in which you can find entertainment and exhilaration in so simple and natural a pastime.—Burroughs, *Pepacton.*

All winter I had been promising myself the pleasure of watching the flowers unfold in the Bogs of Etchowog. On May 25th I reached the old farm on Mount Œta, having departed from New York on May 14th, fully equipped as a bog-trotter, with hunting-boots, rubber gloves, short skirts and vasculum.

My route was through New Haven and Hartford, across the States of Connecticut and Massachusetts. On my way I stopped for a brief visit at the home of a friend in New Haven. In her garden, I found a corner of the Taconic woodlands awakening. Here, in line and on time, stood five modest Yellow Lady's Slippers (*Cypripedium hirsutum*), members of the Orchid Family; while along the same border clusters of the Showy Lady's Slipper (*Cypripedium reginæ*) were pushing their dewy-tipped beaks into light and sunshine. Although rather late in their blossoming, compared with the other sisters of this genus in New England, this species usually reaches its prime about June 20th.

On the east side of the garden towered an ambitious row of ferns, some twenty root clusters or more, including many rare species. Here was an especially queer little strap-like leaf, which one would scarcely call a fern unless one were a professed fern-hunter. It is the rare Walking Leaf (*Camptosorus rhizophyllus*), the scientific name meaning a bent heap, and the appearance of the plant indeed is suggestive of the name. The frond is from four to twelve inches long, springing from a heart-shaped base and reaching out a long, narrow runner, which readily roots at the end again, and thence takes a step onward, and so on, until three or four steps are taken, often in this way forming a beautiful carpet for the cold gray lime rocks, which it prefers in its native haunts.

The Walking Fern is shy in its habitat, seeking the most hidden crevices in ledges along our mountain sides. I have collected it in many dark ravines, as well as along dry, rocky ridges in the Hoosac Highlands. It takes kindly to cultivation for a season or two, and then dies out for want of its natural soil of limestone.

A short walk toward West Rock, New Haven, showed me how far advanced the season really was. Here were crowds of children playing in fields covered with violets and bluets, and farther down in the damp meadows were long, serpentine lines of gold, where the Marsh Marigolds (*Caltha palustris*), known commonly as Cowslips, were already fading. On the edges of the swamp, the Marsh Buttercups of the Crowfoot Family (*Ranunculaceæ*), were lifting their shallow yellow cups to catch the sunshine. We wandered on through a pretty, wild bit of young woodland until we reached the border of a murmuring stream, creeping onward through the vale and meadow, touching the blossoming orchards here and there, and freshening the sweet white violets on its brink.

North Adams, Massachusetts, was to be my next station. This city is about two hundred miles from New York, among the Hoosac Highlands. I almost expected to see reluctant snowdrifts still lingering in the fence corners and shaded pine glens of this part of "Beautiful Berkshire," and I half hoped to find a few late clusters of the Trailing Arbutus (*Epigæa repens*) creeping through the cold, mossy ravines.

Upon my arrival in North Adams, I looked through the bogs under the brow of Hoosac Mountain near Aurora's Lake, and I could perceive scarcely any difference in the progress of flowers or foliage here from that of the region from which I had just departed. Dogwood, apple trees, violets, anemones and wake-robins were in blossom, while in the deeper bogland I found one lone, pale Pink Moccasin-Flower (*Cypripedium acaule*).

American White Hellebore, so commonly known as Indian Poke or Itch Weed (*Veratrum viride*), had already sent out a luxuriant growth of green leaves, which for a moment deceived me—as it had done many times before—by its resemblance to the foliage of the Showy Lady's Slipper. The leaves of both these plants are plicate, and have ever been confused even by the earliest herbalists. Unrolling a spike of leaves one day, I found I had actually disturbed the buds of the queen of the Lady's Slippers instead of the Hellebore, although they proved to be blasted. No doubt some warm day had started them prematurely, frost and cold rains later proving their ruin.

Here on a sheltered damp hillside, I found my first clusters of the season of the Pink Azalea (*Azalea nudiflora*), which is commonly known hereabout as Swamp-Apple, and which is very similar to *Rhodora Canadensis*. These species belong to the Heath Family, one of the largest among the flora of Hoosac Valley. The beautiful pink flowers of the Great Rhododendron, which measure from one to two inches in diameter, render it the most charming species of this group. It is cultivated extensively, but grows in its natural wild state, in this region, only along the margins of ponds near Montpelier and Wells River, in Vermont.

The Botanizing Can, or Vasculum, showing the White-Petaled Lady's Slippers and Maiden-Hair Fern.

The American Mountain Laurel (*Kalmia*), which becomes so gorgeous later in the season, the Lambkill, Labrador Tea, Andromeda and the Cassandra are closely allied species of this group, common to this region. Other familiar members of it are the Trailing Arbutus, Gaultheria, and the Creeping Snowberry. They may be found in Aurora's Swamp.

North Adams is not far from the sources of the south and north branches of the Hoosac River, in a wild and rugged portion of Berkshire. The Hoosac proper is formed at the junction of these two streams, in the vicinity of the Print Works near Marshall Street in the city, and flows on gently in a northwesterly course to join the Hudson, near Lansingburg. Mountain streams in this region are numerous, and flow musically down through deep chasms and over great marble precipices, to swell the Hoosac as it glides slowly out through the deep-cut valley.

"We Hold the Western Gateway," is part of the inscription on the seal of the city of North Adams, which is known as the "Tunnel City." This is practically true, for the sole gateway of the trade from the Western States passes though the flinty wall of the Hoosac Mountain, in order to reach Boston direct. The idea of opening a path for transit through the "Forbidden Mountain," as the Indians called it, was conceived six years after the first mail-coach and four-in-hand rattled through the street of this town to Greenfield, in 1814. It was found impossible to build the projected canal from Boston to Albany. The estimated cost of building the tunnel was less than two million dollars, but when it was completed in 1875, the total financial outlay had amounted to over twenty millions. Until January 1, 1887, this tunnel was owned by the State of Massachusetts, when it was purchased by the Fitchburg Railroad. It is four and three fourths miles long, and twenty-six feet wide, permitting of double tracks. The arch is from twenty-two to twenty-six feet high, and at each portal there is a massive granite facade.

Whenever I come to the Hoosac Valley, I enter, if possible, by way of this tunnel. I seem thus to close away the outer world, and to penetrate a new realm hidden here in the seclusion of the marble highlands. This triumph of man over the power of Nature needs no further introduction here. I can never forget, however, the weary years of hardships endured by those who toiled in its construction, entombed within the heart of the mountain, subject to the dangers of quicksands, falling rocks, damp and gases, explosives, fire and starvation, before the great work was accomplished.

I enjoyed the ridges in the pastures along the foothills of the grim-faced Tunnel Mountain, and about Aurora's Lake, which reflects like a pretty little mirror the rugged beauty of the hills. This lake is partly natural, but now dammed artificially. Every line of its terraced shores bears the scars of antiquity, which would indicate that ten thousand years ago a larger lake slept in this hollow vale which geologists have

Mount Greylock's Brotherhood—the Berkshire Highlands, from Mount Œta, Bennington County, Vermont, Showing the College Town of Williamstown in the Valley.

estimated at a depth of six hundred feet. Here are rich deposits of glacial drift, and northeast of Aurora's Lake are sphagnous swamps, where I find many rare orchids and early spring blossoms. Here both the pink and yellow Moccasin-Flowers bloom in May, while in June the queen of the tribe unfolds her white-petaled purity.

This bogland is very similar to that of the Swamp of Oracles in Pownal, in District Fourteen, save for the openness of the former's shores. Aurora's Swamp is located in a deep flinty basin, surrounded only by mystical age. The name of "Greylock" appears to be derived from the lowering cloud-mist so often capping the whole Brotherhood at early dawn or before a storm.

Vermonters who, from the hills at a great distance to the north, view this group of mountains, depend upon this capping of clouds as a forecast of the weather. Among the old folk, it is known and designated as "Greylock's Nightcap," a portent of a coming storm.

Mount Greylock, the highest swell of this range, is 3600 feet above sea level, and commands a variable and extensive view from its bald summit, on which was early erected that first wooden observatory, during President Griffin's term at Williams College. Here the poet and the philosopher, Hawthorne and Thoreau, have climbed to meditate.

Many a message has gone forth from these heights to bless the busy world. Scarcely is there a son of old Williams who does not recall the mountain-day excursions led by Professor Albert Hopkins, and the glory of old Greylock at dawn and at the sunset hour.

Thoreau writes of it: "It would be no small advantage if every college were thus located at the base of a mountain, as good at least as one well-endowed professorship. It were as well to be educated in the shadow of a mountain as in more classical shades. Some will remember, no doubt, not only that they went to college, but that they went to the mountain. Every visit to its summit would, as it were, generalize the particular information gained below, and subject it to more catholic tests."[1]

The peak especially designated as Saddleback Mountain is at the junction of the eastern abutments of that huge wall of Taconic Brotherhood which appears south of the old battle-ground where formerly stood the early border Fort Massachusetts, on the Harrison flats, near the flag station of Greylock. The union of Mount Williams, sloping to the east, and Prospect Mountain to the west forms the seat of the saddle.

Mount Hopkins—so named in honor of Professor Albert Hopkins of Williams, the first nature-student of our land, making excursions afield in 1833—lies south of these. Old Greylock, proper, lifts its lofty brow still farther south, being situated about in the centre of this great range as it extends from east to west.

Beyond Greylock stretches a long, misty line of blue peaks against the sky, which if observed from Mount Œta at the north, in Bennington County, Vermont, may be traced to the southwest to Symond's Peak, the lowest of the group, named in memory of Captain Symond, who led the volunteer forces from our hills and vales to the memorable Battle of Bennington in 1777. Bald Mountain is also in the vicinity, and the closing in of these several peaks has conspired to form what is known as the "Hopper," and the "Heart of Greylock." The hollow vale amid these heights has the appearance of the hoppers used by millers years ago.

Surely in the heart of the Taconics we are in one of the oldest rock formations of the earth, and the green terraced stairs lead us slowly down to the deep-set valley of the Hoosac, where once slept that ancient lake. All that now remains of that Lake of Dawn is pocketed in the basin under the Hoosac. The shores of Aurora's Lake are lonely and still, save for the marsh thrushes which skim low over the waves and whistle shrilly. The groves of pine to the southeast are the haunts of solitude, and those who wander here can well imagine that the Æolian harps among the whispering trees are repeating a music of ages past, when only wind and waves were known to these hills.

Amid these damp and reedy shores and swampy woods are tail brakes and delicate Maiden-Hair Ferns. Here, too, the tall and stately Royal-Fern (*Osmunda regalis*) flourishes in deep seclusion, sheltered by the low-branching pines along the shore. It grows from two to four feet high in this locality, and is of a deep rich crimson-green tone against the grasses and bushes near. Mounds of moss, marking one of the trees of a primitive forest rotting below the soil, are thickly carpeted with the leaves of the Dog's-Tooth Lily. Indeed, the picturesque paths which lead through these woods wind through a veritable fairy-land of flowers and ferns. One of these trails runs southward through rocky pastures, swamps and thickets, toward the Tunnel's western gate.

Along these slopes, among the limestone rocks, I found rows of the Ebony Spleenwort Fern, rather rare in this much-travelled way; and on the brow of this ridge were many species of common fern. The pastures are barren and dry, with few bushes to break the dreary horizon, as one approaches the western portal of the Tunnel.

I came upon one lone Apple-Thorn bush, of genus *Cratægus* of the Apple Family. Nearly opposite, across the valley of the south branch of the Hoosac, which the Indians named the Ashuilticook, may be distinguished the smoking Limekilns; while still farther southward, the white-spired village of Adams nestles at the base of Greylock, which towers serenely above the shaggy shoulders of Ragged Mountain. I wandered about the edges of the Tunnel cliffs where, in years gone by, had stood the impoverished cabins which sheltered the laborers who tunnelled the Hoosac. I descended into the chasm and seated myself upon the wall of rocks, waiting for the trains to appear and disappear at the portal in the side of the hill. Presently one from the West crept ponderously into the cavern. The echoing roar was smothered, and died slowly away until it became an indistinct murmur. Not long afterward I felt, as well as heard, the low breathings and rumblings of a locomotive coming in the opposite direction. I heard its subterranean groans as of a great spirit, while the smoke poured forth, pushed in volumes before the engine, wreathing and curling about it as it emerged, and partially concealing its grim outlines.

The faithful watchman, a modern Eckhart, sits before the entrance of the Western Gateway of Hoosac Mountain, and warns the people against entering through this portal to the greater world that lies beyond. It is as if he wished to guard these children of the marble highlands from the risks attendant upon the wild whirl of life beyond these quiet hills.

The sun was setting as I left him, calm but alert, at his post of duty, trimming and lighting his colored lanterns for signals of danger or safety to the approaching trains. Climbing up by the path which passes the little red cottage on the crest of the hill on the north bank of the chasm, I returned leisurely homeward, winding over the hillsides, far above Aurora's Lake, then down along the borders of the swamp-lands. In the crevices of rock were creeping colonies of the Common Polypody (*Polypodium vulgare*). Along the edges of this bog are still seen the primeval stumps of the pine and hemlock forests, which clothed these hillsides when only the Redmen dwelt and hunted among these wildernesses.

In May and early June these decaying stumps are usually draped with Painted Trillium and the delicate vines of Gaultheria and the Creeping Snowberry, while the Arbutus trails about luxuriantly, covering up the ruins of years.

Notes:

[1] Thoreau, *Week on the Concord and Merrimack Rivers*, p. 244.

The Western Gateway of Hoosac Mountain, the Entrance to Hoosac Tunnel, North Adams, Massachusetts

2
Ball Brook and the Bogs of Etchowog

Fringing the stream, at every turn
Swung low the waving fronds of fern;
From stony cleft and mossy sod
Pale asters sprang, and golden-rod.
—Whittier, *The Seeking of the Waterfall.*

On May 25th I reached Pownal, Bennington County. Upon the following day I explored the great swamps of Etchowog. Prepared with luncheon, vasculum, basket for roots and my hound Major, I started on one of those happy excursions such as Thoreau recommends we should take, "in the spirit of undying adventure, never to return,—prepared to send back our embalmed hearts only as relics to our desolate kingdoms."[1]

Ball Brook, a sluggish stream flowing northwardly to East Pownal swamps—commonly called the Bogs of Etchowog—has its source in the marshy hillsides northwest of the schoolhouse in District Fourteen. Two streams flow from this valley. One is called Ladd Brook, running southwesterly and following the windings of the shady Ladd Road to Pownal village, where it joins the Hoosac River. The other stream, Ball Brook, flows north and northeast onward through innumerable and unfathomable swamps, to Bennington village, ten miles north, there meeting the Walloomsac River, which is also a tributary of the Hoosac, farther northward in its course. This brook is rich in a continuous chain of peat bogs—rich from an orchid-hunter's point of view.

Although I have been familiar with this region from childhood, viewing it from the roadside only, I never at any time had ventured to follow Ball Brook through all its meanderings to the Bogs of Etchowog near Pownal Pond, a distance of some three miles. This would not be a long walk on a fair road, but it becomes rather dangerous and formidable when leading through quaking marshes in the soaking currents of a stream.

A short distance to the right, north of the schoolhouse in Number Fourteen, there is an old pathway nearly overgrown with bushy pines

Ball Brook, in the Swamp of Oracles, Pownal, Vermont

Here, let these rivulets forever flow!
Drink from these highland domes the melting snow;
Drain from the dark ravines, and hollows near,
The mountain cascades, flowing soft and clear.

G. G. N.

and birch and chestnut underbrush. This I followed, entering the hollow under the brow of the hill, and passing along the wood road which skirts the margins of one of the deepest, darkest jungles in these regions. The old people look upon it as akin to "Witch Hollow," on the Gulf Road near by, and tell strange tales of ghosts, and of some mythological peddler who was swallowed here in the black mud of this ancient tarn, after having been robbed of his fine silks and precious jewels.

Weird, hollow drummings issue and echo through these shaded vales from time to time. Probably they are due, however, to nothing more startling than the alarum of a partridge, or the hoot of the screech-owl; or the creaking and rubbing of partially fallen trees against their supporting brothers, voicing a portent of coming storm. I hear in this woodland seclusion little save the whispering of the winds, the sighing of the pines, and snapping of dead twigs, mingled with the chorus of the thrushes. The first settlers here about interpreted these wood-sounds far differently; then the primeval forests were dense, and the noises were deep and full of mystery, and there was fear of the Redman's war-whoop. As Burroughs writes: "The ancients, like women and children, were not accurate observers. Just at the critical moment their eyes were unsteady, or their fancy, or their credulity, or their impatience got the better of them, so that their science was half fact and half fable. ... They sought to account for such things without stopping to ask, Are they true? Nature was too novel, or else too fearful to them to be deliberately pursued and hunted down."[2]

I stopped on a corduroy bridge to draw on my high-water boots and rubber gloves, for one feels safer when entering this dense swamp if protected from poisonous roots and foliage, biting insects and things that creep and crawl.

I had started out with small belief that I would find any prime blossoms of the Orchid Family, for nothing of importance had yet unfolded in Aurora's Swamp in North Adams. But when I penetrated the heart of these rich, warm glooms, I found waiting for me a fragrant company of Dwarf Yellow Lady's Slipper (*Cypripedium parviflorum*); and innumerable Stemless Pink Lady's Slippers, more frequently called the Indian's Moccasin-Flower (*Cypripedium acaule*), stood as sentinels on the dryer edges of the swamp.

The Marsh Marigolds were here also in their last stages, fading away, but still sufficiently bright; with the late indigo-blue violets, which rear their faces at least a foot high above the dark pools, to carpet the marsh with gold and purple. Poison Ivy cropped out frequently among these graceful orchids,—a beautiful vine, although unfriendly to man.

It is difficult to describe the dense gloom of this bog, closed in on all sides by high rock-bound hills, which are clothed with pine and yellow birch trees, and which in their turn are but foothills to the higher watershed. It seems to have been a receiving basin for the waste and wear of the heights above for thousands of years. Here are fallen trees of every variety common in southwestern Vermont, and these prostrate giants helped to form a safe footing through the quaking bogs.

Many cold springs under the hill to the south conspire to freshen the marsh, and after sluggish oozing northward, they unite and form the brook proper. The stream leads directly through the heart of the swamp, and at last, gathering force, rushes down over rocky slopes, presently to enter another swamp of greater breadth, filled with different trees and flowers.

The Showy Lady's Slipper (*Cypripedium reginæ*) was just sending forth its tiny roll of leaves, so I could not expect prime blossoms before June 15th at the earliest.

Seated on a decaying log, I ate my luncheon, with Major before me begging impolitely for his portion, until I divided my cake with him. The mosquitoes were so troublesome that I decided to push onward. Carefully picking my way out of the swamp, I crossed the muddy brook, and found myself in a dry, rocky pathway which winds around the hillside, but still keeps within sound of the brook's murmur.

In exquisite little glens beside the path were Painted Trilliums and Stars-of-Bethlehem, while the white and gold stars of the dainty Goldthread (*Coptis trifolia*) were shining amid the moss and their own glossy green leaves.

In the bend of the stream a little farther on were some of the most graceful little ferns, just near enough to the brink to catch now and then a dash of spray from the rushing waters, swayed in the coolness all day long, adding beauty to the nook.

Still farther on, I saw that by crossing the stream I could enter a little ravine to the right, which promised hidden treasures. I waded through the brook, which was too wide to jump across; I found that it was also rather too deep for my boots, and that there were very few stepping-stones to make a dry crossing possible. But of what matter is a little water in one's boots, when seeking the Gardens of the Gods? I landed safely on the opposite bank, after frightening many a shy, speckled trout from his hiding-place in this ideal fishing-hole.

I was now in a small, low-lying glen where foot of man has seldom been. The soil, though much drier than the ground over which I had recently passed, displayed a honeycombed appearance, showing where the water had oozed away through the rich leaf-mould to seek the flowing stream beyond.

Whole constellations of star-flowers were here; both the Painted and Crimson or Nodding Trilliums were abundant, asserting themselves and their rights, if size of flowers and leaves may indicate strength, among the tall, rank growth of the Common Brake (*Pteris aquilina*), which frequently rise five feet in height. Close by their long, harsh lobes grew the plicate leaves of the Indian Poke or White Hellebore. Skunk Cabbage (*Spathyema fœtida*), so frequent in the swamps along Bronx River in Greater New York, is rarely seen here, although I find lone specimens now and then in Aurora's Swamp in northern Berkshire, and in this jungle. Lily leaves and Dwarf Cornel peeped out from every shadow. Here I found the red-spotted leaves of Dog's-Tooth Lily (*Erythronium Americanum*) and Clintonia (*Clintonia borealis*), as well as the delicate leaves of the False Lily-of-the-Valley (*Unifolium Canadense*), and several species of Solomon's Seal, while the weird Indian Cucumber (*Medeola Virginiana*) rose up everywhere beneath the luxuriant ferns.

Dwarf Cornel, or Bunch Berry, locally known as Bear Berry (*Cornus Canadensis*) was about to set its fruit. These berries are of a deep vermilion color, and eatable if one has the patience to sever the seeds. From the bark of this species of the Dogwood Family is extracted a tonic which is very bitter.

I found the beautiful Star-Flowered Solomon's Seal (*Vagnera stellata*), and the deeper bogs revealed specimens of the rarer bog species, *Vagnera trifolia*, which, in spite of its name, produced plants with more than three leaves, and many beautiful fragrant flowers of a waxy white color. Indian Turnip (*Arisæma triphyllum*), more commonly known to-day as Jack-in-thePulpit, was numberless; the little priests in the pulpits were dressed in cardinal's robes trimmed with stripes of green, white, and purple.

This sylvan retreat which yielded so many specimens of beautiful flowers I called the "Glen of Comus," for I could not rid my thoughts of the deep, dark woodlands where Sabrina was lost among the enchanters.[3] I fancied that the Purple Trilliums stood with nodding petals bowed down to earth as though they were guilty of some crimson sin and dared not lift their faces to the sun.

I gathered from every species some perfect treasure, and then returned, wandering once more beside the cool brook. I wondered if it carried all the memories of the forest fastness, gleaned among the roots of our frail, beautiful hillside flowers, through the mighty rivers to the deep seaweeds and strange aquatic blossoms which had at one time bloomed among these very hills ages and ages ago.

Climbing a fence, I found myself in a parched, short-cropped cow-pasture, but the stream soon passed into a large tamarack swamp, where in many places neither man nor beast can wander with ease or safety. I rested under a wide-spreading pine tree, looking the marsh over to choose the best path through it, for I still had some distance to walk before I could reach Pownal Pond and the Bogs of Etchowog.

In order to make my journey less burdensome, I decided to leave my treasures of gold and crimson hidden in this stream, where they would not only keep fresh, but would be much safer than with me. I felt that they would be reasonably safe from marauders, for orchids are far more numerous than human beings in this forlorn locality; for where verdant meadows might spread were only uncultivated, almost impassable, dismal swamp-lands.

The Showy Lady's Slipper—The Queen of the Indian's Moccasin-Flowers.
(*Cypripedium reginæ.*)

Few poets have ever sung the praises of the Queen of the Moccasin-Flowers, although a lovelier flower never beckoned to poetic fancy.

At last my flowers were safely placed in the bend of the brook near an old pine stump, where I made them fast, covering them with the coarse brakes which grow everywhere; then I strode on northward through the tamarack swamp. This marsh covers a large part of Ball Farm, from which the brook crossing it derives its name.

Through the trees I could see the old weather-worn farm buildings, nestling in the shade of a dozen or more large, thrifty maples, and now and then I heard a faint murmur of distant voices. Suddenly they subsided, and a small dog's shrill bark told me that I was discovered, mistaken perhaps for the veritable "Witch of the Hollow," by the present colored occupants.

There was no use in trying to follow the stream now, for its windings were intricate and indefinite. It wandered all over the meadow marsh, and splashed out in one great mud-hole, similar to that of the jungle in District Fourteen, save that the meadow here was open, with very little low tangle or underbrush in sight. Innumerable tamarack trees, however, lifted their graceful spires throughout the bog; yet this did not prevent the meadow from appearing flooded with sunshine.

Away over on the west side of this swamp were many low-spreading trees of virgin pine, contrasting prettily with the lighter greens of the delicate spires of tamarack. Between myself and the shore on either side of this mud-swamp waved acres of Fleur-de-lis, which would soon color the whole meadow with royal purple. Still westward of this lay an alder swamp. This shrub, called Speckled or Hoary Alder, belongs to the Willow Family, and grows about fifteen feet high, along swamp meadows, forming dense thickets.

Many saucy swamp birds dwell here and appear tame; they came chattering after me, fearing, no doubt, that I might be in search of their nests and birdlings.

Under the pines on the border of the swamp I rested, finding the while tender young Wintergreens (*Gaultheria*), and many edible red berries, called Checkerberries, fruit of *Gaultheria*, sometimes known as Partridge-berry and Boxberry. The last two names are more frequently applied to the fruit of *Mitchella repens*, found growing in company with *Gaultheria*, and producing edible scarlet berries on a trailing vine, resembling myrtle. The flowers of this vine were now in bloom, giving forth a delicate perfume. Their white and pinkish-purple blossoms dotted the moss with a brilliancy like that of the Trailing Arbutus (*Epigæa repens*), so lately faded.

The buds of *Moneses uniflora* were putting forth their "single-delight," the name coming from their solitary flower. Here also were quantities of the glossy, waxen leaves of Pipsissewa or Prince's Pine (*Chimaphila*), and low creeping evergreens. Common Club-Moss and Ground-Pine were interlaced in their dark green beds, where had recently nestled the clusters of arbutus, now brown and faded, although the mossy hummocks still held the fragrance of their luxuriant green leaves. Whittier, writing of these spicy flowers, associated them as the first flowers which the Pilgrims looked upon after their landing on the bleak shores of New England, at Plymouth, in the spring of 1621, and says:

> Yet, "God be praise!" the Pilgrim said,
> Who saw the blossoms peer
> Above the brown leaves, dry and dead,
> "Behold our Mayflower here!"[4]

In New England the Arbutus is commonly called "Mayflower,"—not that it blooms especially in the month of May, for it has been found in northern Berkshire as early as February and March. My observation is that prime blossoms are found in the Hoosac Valley region from March 15th until May 15th. I have also gathered beautiful clusters as late as June 23d, in cold nooks beneath the shades of spruce and pines. Their spicy perfume is ever the delight of New Englanders.

Scrambling with difficulty over a fence which sagged toward me, I entered a neighboring pasture, finding here more alder trees. Small tamaracks, Christmas spires of spruce, and pine seedlings filled the pasture with fresh evergreens, making me fancy myself in a cultivated park, so regular and trim they stood. Eastward crept Ball Brook, wandering through deep, reedy grasses, where here and there stood tall spikes of last year's Cat-tail Flag (*Typha*). Here also grows the Sweet Flag or Calamus (*Acorus*), which is not only good to eat, but a panacea for sore eyes. The cat-tails stood stiffly erect, as if guarding the blossoming bog, and serving, notwithstanding their dignity, as perches for the saucy finches which still chattered after me.

Now I passed through a barway to the right, ever in hearing of the gurgling stream, which had reached a hard, dry, gravelly soil, abruptly following the downward slope around a hillside. A well-worn sheep path led me down into a bog similar to the Glen of Comus in District Fourteen, only if anything more wild and weird. Through the openings between the trees and knob-like glacial hills, I caught glimpses of the bold, rugged form of the Dome, standing coldly against the eastern horizon.

A glance through these glooms revealed another colony of the Showy or White-petalled Lady's Slippers just bursting forth from the earth, perhaps four inches high. I have found them frequently in these bogs, when full-grown, standing three feet tall, but the usual height is about two feet; and in open meadow swamps often only eighteen inches, owing to the crowded soil, choked with grasses and low shrubs. In about three weeks these bogs would be gay with dainty Moccasin-Flowers.

In the upper part of this swamp, I found a rather quaking corner devoted entirely to the deep green leaves and tall, white-bearded spikes of the not common Buckbean (*Menyanthes trifoliata*), a distant cousin of the Blue Fringed Gentian. I know of several colonies of this rare plant in the bogs hereabout, where it grows plentifully, in its pet localities. It is liable to grow ever undisturbed, I am sure, since it chooses such dangerous swamps in which to flourish.

Thoreau mentions that Hodge the geologist once found at least an acre of this species. He writes: "We reached Shad Pond, or Noliseemack, an expansion of the river. Hodge, the assistant State Geologist, who passed through this region on the 25th of June, 1837, says, 'We pushed our boats through an acre or more of buck-beans, which had taken root at the bottom, and bloomed above the surface in the greatest profusion and beauty.'"[5]

After leaving this jungle,—which reminded me of the luxuriant vegetation of tropical swamps,—I pushed onward, ever nearing the broad marsh-lands of Etchowog, east of Pownal Pond, in the shadow of the Dome. A shaded wood-road winds around the base of the hill, through an open gateway into a thrifty, well-kept apple orchard. This adjoins the old Kimball homestead, and I therefore designated these marshes Kimball Bogs. Out through the orchard meadow I passed, crossing the dusty highway which leads northward around the pond. There are several roads leading to Bennington village; some are rough, some are narrow and hilly, while others are broad and easy. The one to the left, called the Middle Road, follows through Pownal Centre to the county seat of courts and justice. By keeping to the right, one arrives at the same destination by the rough but picturesque East Road, under the brow of the Green Mountains. A direct route from Pownal Pond to Bennington is by way of the Hill Road, which leads directly north between the other highways. Thus the region is intersected from east to west by many roads running northward. I invariably recommend the Hill Road to the traveller who enjoys beauty of landscape. On this way, if he be a keen observer of nature, he will find much pleasure.

Instead of going by the trodden way to Pownal Pond, I chose to follow closely the windings of Ball Brook, which at this point of the road, opposite Kimball's barns, mingles with another mountain torrent that comes down from the spring heads above Thompson's Pond, under the Majestic Dome. The main current of this stream continues with the bend of the road, taking with it the volume of the water of Ball Brook as it crosses the greater stream. The courses of both streams are unnatural, having been removed, over one hundred years ago, from their original channels in order to form a mill-pond for sawmill use. Originally, I am told, a dense forest of pine trees occupied the hollow where now the waves of Pownal Pond wash over the decaying stumps.

The natural lake bed lies in these broad, sphagnous meadows east of Kimball's homestead, winding around to the north, where now wave various small shrubs and trees. Barber's sawmill, which stands close by the roadside, east of the pond to-day, is slowly crumbling away for want of use. Water finds its level, and although forced to go by the

The Fleur-de-Lis. (*Iris versicolor.*)

The Fountain of Arethusa, near the Bogs of Etchowog, Pownal, Vermont.

"If you would get exercise, go in search of the springs of life. Think of a man swinging dumb-bells for his health, when those springs are bubbling up in far-off pastures unsought by him!" —Thoreau

roadside, Ball Brook still seeks in part its old channels through the ancient meadows of Kimball's Farm, where the stream is silent and elusive, as it glides among the tall, lush grasses. Walking along the borders of this hidden brook, through the tangle mingled with daisies and buttercups, I lost the stream entirely, only a line of gold marking its sleepy wanderings,—for marsh marigolds were still plentiful here, ever following the edges of the brook.

Hellebore grew over the swamp, and the tall grasses took on coarser forms as I waded farther on, deeper and deeper into the sphagnous grave of the ancient lake. At times it seemed so soft and spongy that I questioned my safety, even doubting the possibility of a search party securing my "embalmed heart," if once I became fast in the mud, so I began to edge up toward firmer ground and the rocky hills near by.

This was the most uncertain swamp I had ever traversed, and not quite safe for one to wade through alone. It is reputed to have been at one time the bed of a great lake, as evidenced by the terraced hillsides about it. Its waters might still linger beneath the black-peat and forest debris which support the trees and spongy sphagnum. However, a fence closed off the most dangerous parts of the bog to keep back the cows from the mire and "dead holes," as the unfathomable places are designated by the lads who penetrate these bogs for the marsh cranberries in the autumn.

I searched through this meadow for the Large Purple-Fringed Orchis (*Habenaria grandiflora*), thinking perhaps I might find the leaves, although I was somewhat too early to secure the flowers, since they are not due until June 20th and later.

On striking out for the hillside path, I found many problems to solve. It appeared impossible to gain a firm or safe footing in the sphagnum and mud, so securing a fence board which had been hurled about the marsh by the winds and storms, I slapped it down upon the soft earth and moss, and walked its length of eight feet. Then quickly relaying it, while my feet sank lower and lower in the moss, I hastened to pull out my muddy footgear and walked the length of my bridge once more,—repeating this perilous feat several times, until I had finally crossed the "dead hole" and stood on *terra firma* once more.

There is certainly no experience like being stuck in a bog to arouse fearful forebodings. The discouraging effort to keep one foot above the ground only to find the other sinking deeper is most terrifying, and leads to hasty and excited movements which but increase the danger, and may finally lodge both feet fast in the mud. In such a case the sight of a board fence upon which an elbow may be rested is as welcome as a sail to a ship-wrecked mariner. There is in truth much art and science in walking safely through mud and sphagnum. One cannot saunter over the surface, and meditate at ease, but one must be ever alert, elastic as a rubber ball, and quick to feel a danger before it can be seen.

The fields and woods are a good deal like the books we read: the more we become familiar with printed page or forest path, the oftener we return to certain thoughts and trails that lead us back to scenes and associations enjoyed before. I like to mark passages in books I love, here and there, as I would blaze a tree to guide me to the haunt of a cool stream or a rare flower's hiding-place. Whenever I turn to such passages, I find that time and season have expanded some new thought in my mind, even as they have developed the buds to full-grown flowers since my first journey through the wood.

There is a beautiful cold spring under the hill near the swamps of Etchowog. I have known of it all my life, and were I to visit this region every day for months, I should invariably be drawn unconsciously to this fountain. It is here that I quench my thirst and rest after wading through the neighboring swamps. I have turned many stones here in the past, and lifted the dead leaves from the choking throat of the spring. I have gathered the sundew growing in the moss fringing the banks; and in the sweet solitude and peace I have dreamed many dreams, inextricably mingled with the music of the stream.

To-day I sought this spring to rest. I bathed my face and combed my hair over Nature's own mirror, after taking a generous draught from the sparkling water. It bubbles and gashes continuously from under the rocky hillside, bringing sand and delicate-hued pebbles to scatter in the bottom of its bowl the year round. I rested here a full hour, and rinsed the mud off my boots.

From here it is but a short walk to Barber's Mill at the foot of Pownal Pond. Alders, willows, shad-bushes and pink azaleas, small white birches, tamaracks, pines, and beautiful swamp or soft maples fill the broad expanse of marsh-land to the right; while the rocky, burnt-over, and blackened hillside rises up to the left. I was tempted into the deeper underbrush, but proceeded very slowly, as the treacherous bog was so spongy with sphagnum that I would often sink from twelve to fifteen inches into its soft, pink depths. But here I felt secure, since there were many fallen trees and growing saplings to which I could hold and cling, in case I stepped into a "dead hole."

Here, half buried in the moss, I found hundreds of crimson-veined Pitcher Plants, or Side-Saddle Flowers (*Sarracenia purpurea*), which bear olive-green, purple-veined, vase-like leaves that hold rain and dew. Often the species varies in color, and its absolute greenish-yellow with lighter green veinings. Many of the larger pitchers hold fully a tumbler of fluid. Their brilliant-hued brims are edged with crimson ridges, delicately coated with honey, thus enticing flies and moths to drink from the nectar beyond the brim. The more common prisoners are small flies and moths, but one day I found two dozen snails captive in the larger leaves of an ancient plant, for if once within, there is no escape even for snails. Consequently the Pitcher Plants—locally called St. Jacob's Dippers and Dumb Watches by the children—are considered carnivorous plants, since they are flesh-eating by nature.

Round-Leaved Sundew. (*Drosera rotundifolia.*)

This is also true of the small Round-leaved Sundew (*Drosera rotundifolia*).

These plants are traps that not only cunningly entice, but actually entrap and slowly devour their victims. Sundew delights in being fed beefsteak, and Professor Bailey cites Darwin's experiment of feeding them steak, which "they accepted as readily as an insect."[6] The Sundew is plentiful in these mossy bogs. It has red and white, dewy, bristling, round leaves, with long petioles spreading in a tuft. When a small fly or ant touches these sticky bristles or tentacles on the upper face of the leaf, the points of the outer row slowly turn inward, holding their prey closely until it is dead.

Like the enticing honey of the Pitcher Plant, the viscid fluid of the Sundew attracts the flies, and, once alighted upon it, they become entangled and doomed to certain death. After drawing the juices from their victim or bits of steak, they relax and slowly regain their normal positions. The glands of these leaves send out drops of a clean, sticky fluid which glitter like dew drops in the sunlight. The plant sends up a short spike of insignificant, whitish-green, bud-like flowers, which are said to open briefly one by one in their turn, each morning in the sunshine, till the whole spike has unfolded. Each flower turns brown and fades before the successive bud unfolds, so that there is never more than one full-grown flower to be seen at a time. This is not the case with the flowers of the Pitcher Plant. I found many crimson, ball-like buds sleeping tucked up in their mossy beds. They would be in their prime in a week or ten days.

Here I discovered some fine specimens of the Pink Moccasin-Flower, and I was just about to pluck one, when behold—stretching at full length, basking in the sunshine on one of those sphagnous stump mounds, lay a snake, very near the coveted blossom. He may have been black or he may have been checkered or variegated and even charming and beautiful to the snake-hunter, but to the orchid-hunter he was not a prize worthy of a place in the vasculum. I did not wait to study or designate him or count his diamonds, but softly stole away, leaving him still cunningly sleeping, in waiting for prey, beside that gorgeous Moccasin-Flower.

I now regarded with suspicion all the holes in the soft mounds of moss, as the possible homes of snakes, that might object to visitors in their Eden. Immense ant-hills were numerous, and the occupants may have afforded food for Satan's prototype in his idle hours. Now and then the drum of a frightened partridge, giving her alarum, assured me that her brood of chickens was hidden under the leaves and logs not far distant. It is very probable that snakes in these bogs devour small birds and frogs, and lie in wait for them, as I found the one that I had seen this morning.

Before continuing my search, I secured a hardwood staff, feeling safer with a cudgel of some kind in my hand, in case I met Satan face to face. In my tussle to sever the birch limb from the green tree, I

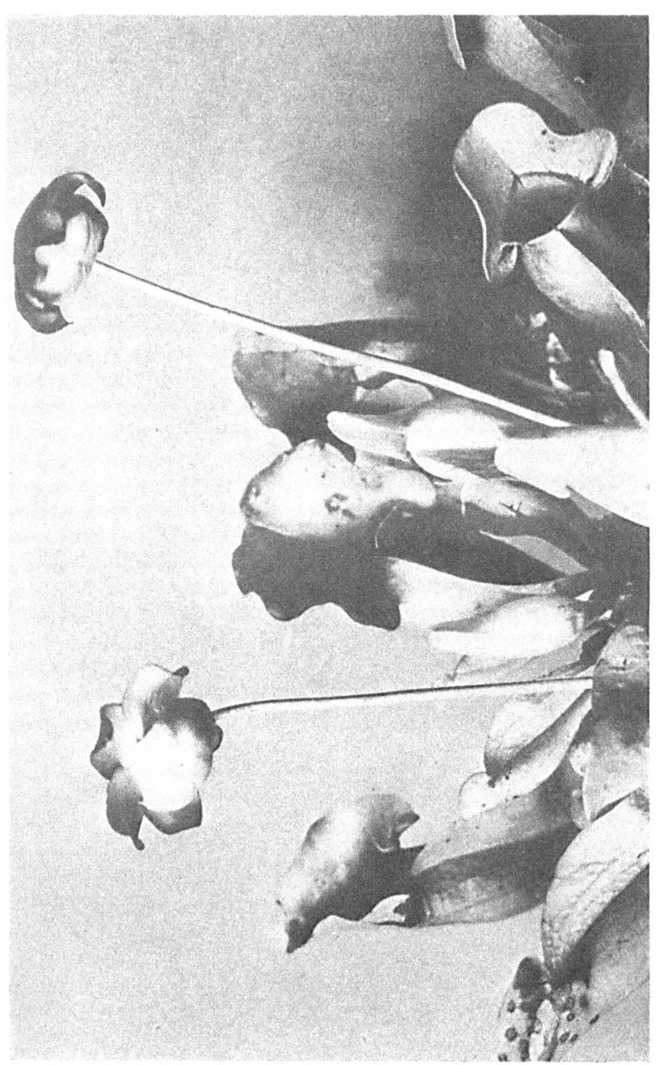

The Carnivorous Plants, commonly called Pitcher Plants, and Dumb Watches. (*Sarracenia purpurea.*)

What's this I hear
About the new carnivora?
Can little plants
Eat bugs and ants
And gnats and flies?—
A sort of retrograding
Surely the fare
Of flowers is air,
Or sunshine sweet;
They should n't eat,
Or do aught so degrading.

—Anonymous

snapped off all the Venus Slippers that I had actually gathered here. I was therefore no richer in actual specimens upon my departure from the swamp than when I entered it; but I carried away memories of that vast solitude and slumbering desolation where foot of man, I dare say, has seldom if ever been.

Now well out of this swamp, I found myself on the edge of an apple orchard, filled with rosy bloom and the fragrance of happy May. A newly planted garden bore witness to human life, and the long rows of potato-hills spoke of industry. Passing through the gate, I entered the East Pownal Road near the mill, and walking down the bank to the right, just north of the mill, where cobblestones had been dumped from the fields, I picked my way into the open Bogs of Etchowog, which lie directly east of the pond.

I wandered up and down through this swamp, finding hundreds of Pitcher Plants, which had begun to nod their crimson buds. Clusters of the Showy Lady's Slippers were springing up on the higher, drier mounds among the lily leaves of *Clintonia borealis* and Dog's Tooth. Fleur-de-lis grew everywhere, while the Poison Ivy flaunted its three-fingered palm on every side. Poison Sumach or Poison Dogwood, sometimes known as Poison Elder, grows luxuriantly in this swamp, and susceptible people have been poisoned merely by passing above along the roadside. By wearing high hunting-boots and rubber or chamois gloves, however, I am perfectly safe in such places. In fact, I never think of these plants as poisonous when brushing through the tangles of bushes and blossoming vines. These species of *Rhus* are in blossom most of the summer. The juice of the plant is resinous, and the fruit consists of white or dun-colored berries.

Going back to the roadside to rest, I took out my color-box and attempted to sketch the swamp I had just left. Eastward, rising boldly in the background, towered the Majestic Dome against the sky. In the middle distance, a long line of alders and willow shrubs blended softly into the blues, here and there dashed with the crimson and gold swamp-maple buds; while still nearer, amid the low, grassy reeds and poison sumachs of the wet swamp, three tall, stately pines reared their shaggy green forms against the dark blue tones of the mountains, lending strength and balance to the scene.

My day nearly spent, I packed away my colors, and started on my return trip, leaving the mill at the bend of the road at three o'clock. Just above the Kimball Farm, I came to a pent-road leading through the pastures to Ball Brook Farm, where I must go to get my Moccasin-Flowers, left hidden in the stream. I found them as fresh and fragrant as if just gathered.

The walking was good, so I exchanged my high, heavy boots for low shoes, which were much more comfortable for dry paths and climbing hillside roads.

Going directly up through the cow-pastures along the border of the Glen of Comus, I came upon a colony of Pink Moccasin-Flowers,

The Bogs of Etchowog, Showing the Dome in the Distance, Pownal, Vermont.

"There are not only stately pines, but fragile flowers, like the orchises, commonly described as too delicate for cultivation, which derive their nutriment from the crudest mass of peat. These remind us, that, not only for strength, but for beauty, the poet must, from time to time, travel the logger's path and the Indian's trail ... far in the recesses of the wilderness."—Thoreau

growing on a sloping hillside under low-spreading pines and birches. Although the spot was shaded, many flowers were unfolding, but they were not so deep in color as time and sunshine would paint them. I counted at least two hundred buds and blossoms, thinking what a feast for the eyes I should have another day, when they were in their prime.

Later, as I turned into the Centre Road, I met Lorenna, one of the school children in District Fourteen. She, too, had her hands full of flowers. I asked her to keep a lookout for strange, small Moccasin-Flowers, hoping thereby that she might find the rare little Ram's-Head (*Cypripedium arietinum*), for which I have so hopefully searched these woods in vain. I had found thus far all the representative species of the Moccasin-Flowers of this State, save the rarer Ram's-Head.

The name Ram's-Head arose from the resemblance of this flower to that of a sheep's or ram's head, the conical or pouched-shaped shoe serving in certain positions to remind the early Canadian children of the noses of frisky lambs' heads, while the twistings of both sepals and petals answered for the ram's horns. This rare species was first collected in Canada near Montreal before 1808. In that year it was transplanted to English gardens by Messrs. Chandler and Buckingham, where they had opportunity to study it closely. For some time it was known as Chandler's *Cypripedium*. Finally, Mr. Robert Brown of England published a description of North American Orchids in Aiton's *Catalogue of Plants*, in 1813, and he must have learned what the children first named it in Canada and Vermont, for he gave it the Latin name, *Cypripedium arietinum*, which it has ever since borne in the science. *Arietinum* signifies shaped like a ram's head, and so one readily observes how the common names of plants suggest to the botanist the origin of the strange Latin names, which are in one sense but the explanations of the common names.

I told Lorenna the story of this stray lamb, and she was as eager to find its trail as I was. The plant is shy at best, the flowers being of the most inconspicuous purple and white shades, found in cedar swamps and on the drier hillsides in mixed wood, of pine, chestnut, and birch. Truth to tell, I was not familiar with the appearance of the plant, nor did I know at what date to search for the blossoms.

After leaving Lorenna, I followed the road homeward, reaching Mount Œta at six o'clock, somewhat dusty and ragged and tired. Old Bonny and the buggy were now suggested as assistants in my trips, when the folk observed my load of herbs and flowers. But bog-trotting in a buggy is certainly beyond the limits of my imagination. It did, however, at that tired moment seem a favorable project, for Bonny and the buggy could wait for me by the roadside while I plunged into the marshes to secure my treasures.

It is true, as Thoreau writes: "we are but faint-hearted crusaders, even the walkers, nowadays, who undertake no persevering, never-ending enterprises. Our expeditions are but tours, and come round again at evening to the old hearth-side, from which we set out. Half the walk is but retracing our steps."[7]

Notes:

[1] Thoreau, "Walking," *Excursions*, p. 252.
[2] Burroughs, *A Year in the Fields.*
[3] Milton, *Comus.*
[4] Whittier, *The Mayflowers.*
[5] Thoreau, *The Maine Woods*, p. 34.
[6] L. H. Bailey, Jr., *Talks Afield*, p. 128. 1885.
[7] Thoreau, "Walking," *Excursions*, p. 252.

3
The Haunts of the Ram's-Head Moccasin-Flowers

I call the old time back: I bring my lay
In tender memory of the summer day
When, where our native river lapsed away,
We dreamed it over, while the thrushes made
Songs of their own, and the great pine-trees laid
On warm noonlights the masses of their shade.
—Whittier, *Mabel Martin.*

The following morning, after my strenuous excursion through the swamps of Etchowog, I was somewhat tired and stiffened, but still ready for a journey which must be made to North Adams, a distance of ten miles from Mount Œta. As it was Saturday, Lorenna's mother would soon be passing over the hill on her way to that city, with butter and eggs, so I decided to accompany her. Lorenna's mother, formerly a teacher in District Fourteen in the neighborhood, had always considered my propensity for tramping through these bogs and woodlands, searching for flowers, as rather "queer." This habit, coupled with my fondness for the poets, led her to believe I had sustained some great sorrow,—perhaps the loss of a lover,—and in those early days she invariably eyed me closely through her green goggles as I met her on the road. My evident annoyance and embarrassment under this scrutiny probably confirmed her suspicions. Nevertheless, she so far forgot her interest in this subject as to tell me to-day that Lorenna, on her way home with the cows the night previous, had found one of the strangest little flowers. None of them had ever seen the blossom before, nor did they know its name. She felt sure, however, that it belonged to the Nervine Family,—as they locally call the Moccasin-Flowers in many New England towns,—from the leaves and the little shoe-shaped flower.

That evening, as soon as the sun sank in the west, and the cool hours of twilight came, I sought Lorenna's house in the vale below Mount Œta. As I sauntered through the fields, the distant sounds of Pownal's church bells and the barking of dogs and the rolling wheels of the home-coming farmers' wagons arose from the valley.

Under my arm I had tucked Baldwin's *Orchids of New England*, a book which I had drawn from the North Adams Library, with permission to keep it as long as I desired, the calls for such books being very infrequent. This work contains many illustrations of species of orchids found in the New England States, and more especially in Vermont, the author having made his excursions and collections of species near Burlington, in the northwestern portion of the State. Among the sketches is one of the Ram's-Head Cypripedium,[1] the species having been collected by him in cedar wood, in the neighborhood of Burlington, where he reports a colony of twenty plants.[2]

Arriving at Lorenna's home, my hopes were realized, and I was introduced to the first fresh specimen I had ever seen of the *Cypripedium arietinum*. Later I was shown the spot where the flower grew. I was hoping to find several plants, but was disappointed. I studied the soil and locality, however, which gave me the clue for fresh trails. We had followed a winding wood-road that led from the Centre Road into the deep pine forests on the Amidon Farm, where the ground was strewn with piny needles and glittering with the Stars-of-Bethlehem, Goldthread blossoms, and the Painted Wake Robins. The broken stem that had borne the conical shoe stood on a rocky hillside, at the base of a chestnut tree. A dwarfed pine seedling was also struggling to grow in the hard soil, among the fibrous roots of the Ram's-Head. The two had probably taken root there at the same time. We marked the spot, and sheltered the plant from the browsings of cows, by planting dead twigs near it.

Before the evening was ended, Lorenna's mother had discovered that others besides myself must have made excursions afield and abog for flowers and herbs, and no doubt at some time in their lives must have also read poetry and made sketches. She became very much in earnest over a text-book on botany, and desired Lorenna to have a child's manual.

Baldwin writes of the Ram's-Head Cypripedium:

"In Northern New England, one is sometimes fortunate enough to gather with the Yellow Lady's Slippers, especially with the dwarf species, the Ram's-Head Lady's Slipper (*Cypripedium arietinum*), the rarest species North America produces, and to me, the most attractive."[3]

The flower is peculiarly conical in shape and slightly fragrant. Baldwin was the first botanist to discover a "musk-like odor" to the roots of this plant, which I also have observed. The structure of this species differs from all other known Cypripediums by producing six distinct parts to its perianth, all the sepals being free to the base. There is in the regular structure of Cypripediums a union of the two lower sepals, usually showing a bifid condition at the apex, when not perfectly united, as shown, if closely studied, in some of the accompanying illustrations.

The brown-pink sepals of the Ram's-Head are all free, and, twisting gracefully, remind one of the horns of a sheep's or ram's head,

The Ram's-Head Lady's Slipper. (*Cypripedium arietinum.*)

In different positions this flower suggests a ram's head.

while the apex of the labellum serves for the nose. The labellum is of a dull purplish color, mottled or checked with white veins upon the crest of the shoe. The apex or toe is of a dull brownish green, the orifice of the labellum is triangular, filled with downy white hairs, and not large enough to admit a baby's finger-tip. The flower, however, varies, as does also the plant, in size, according to the soil and the age of plant, those found in damp cedar swamps being a foot or more in height, adorned with large flowers, while those along the hillsides are from six to ten inches high.

This rare orchid is seldom, if ever, collected by botanists. It is one of the smallest Moccasin-Flowers found in the Northern Atlantic Region. The pigmy of the genus is *Cypripedium fasciculatum*, found under young Conifers in open woods in the swamp regions of northern California, along the Pacific slope, exclusively west of the Continental Divide. The *Cypripedia* found in the Pacific Region are very different from those of the Atlantic, *Cypripedium Californicum*, for instance, producing a simple raceme bearing from three to twelve flowers, all emerging from the axils of leafy bracts, the stem often growing four feet high. The shoe-shaped flowers resemble miniature blossoms of our eastern *Cypripedium reginæ* in color and structure of sepals and petals.

The Ram's-Head Cypripedium is certainly one of the rarest species on the continent, and appears to be more plentiful, if this word can be used of so scarce a flower, in the State of Vermont than in any other region that has been reported in its continental range. It grows in low, damp marl and peat swamps.

Notes:

[1] Henry Baldwin, *Orchids of New England*, Plate 8, 1894.
[2] Henry Baldwin, *Orchids of New England*, p. 38, 1894.
[3] Henry Baldwin's *Orchids of New England*, p. 37, 1884.

4
The Stolen Moccasins

Woodlands, green and gay with dew,
Here, to-day, I pledge anew
All the love I gave to you.
—Alice Cary, *A Lesson*.

Whether the season is premature or backward, the Moccasin-Flowers always appear at the same date, along with the Painted and Crimson Trilliums, in the warm Glen of Comus. I am sure of finding these flowers unfolding, the week previous to Decoration Day, from the 20th to the 28th of May.

On the 30th of May, four days after I had discovered the famous two hundred Pink Moccasin buds on the hillside above the Glen of Comus, I imagined now that they must be in full array, wearing the rich hues of magenta and all the delicate tints of green, white, and pink. When once fully unfolded they change color very rapidly. Late in the afternoon I entered the edge of the Swamp of Oracles in District Fourteen, north of the schoolhouse. My hound was my sole companion, and I heard him in the distance making friends with children, whose voices came echoing from the direction of my fairy-land of Moccasins. A fore-boding that all its beauty had been plundered took possession of me, for I knew that children are instinctively selfish about flowers, and pluck every blossom they see, even though they may throw them away afterward.

I picked my way carefully through the deeper swamp, around in the opposite direction, avoiding thus the children whom I heard approaching by way of the path, their arms laden, no doubt, with the blossoms I sought a sight of. Later my worst surmise was confirmed. Not one Moccasin hung on its stem to tell the tale of the invasion. Here and there were strewn bruised leaves, and stemless blossoms, prostrate on the hillside. I was sorely disappointed, and I exclaimed aloud to the echoing wood that it was a sin,—this stealing all the flowers and leaving none to mature and develop their seed pods for the continuance of the species to be enjoyed by future generations. "And if I ever get hold of these youngsters," I cried, "I'll tell them why!"

The "youngsters" happened to be cousins of mine who had caught the orchid mania from me, and what to them had always appeared ordinary Indian Moccasins, or Lady's Slippers, had now an added value and charm, since they were understood to belong to the Orchid Family. The very hint that I valued them caused strife among these children, eager to show me how many they also could gather in a day. As such treasures, they gathered them, hurrying homeward to tell we how many rare and beautiful orchids they had found. They wondered if I had been near the jungle, as they saw Major, my hound, during the afternoon. I admired their blossoms, now drooping and wilted and sadly bruised, but I never told them just where I had been, nor what I had missed. I had not the actual courage to scold them, since I had set the example for them, but although I find many flowers, I gather at random for mere pleasure very few. Indeed, there is no pleasure in making desolate these choice and hidden retreats of Nature.

There are laws protecting the deer in the Green Mountains and the brook trout in their spawning season, but as yet there is no legal or moral protection to shield the flowering and fruiting season of rare flowers, especially orchids, so scarce in northern New England. Some of our orchids are already so rare, that in localities where, only a few years ago, I found them abundant, to-day hardly a trace of them remains. They have suffered from school children and commerce alike. People seek them selfishly for pleasure and study, while the drug trade demands many roots, and places fair value upon them as an inducement to collectors. These roots are used for infusions, tinctures, and ointments,—a primitive Indian custom and one which, if continued on the present scale, must in time necessarily cease, through extinction of the rarer and most showy species of our native orchids.

The country folk know the Lady's Slippers of genus *Cypripedium* as the Nervine Family, valuing them as a nerve tonic. I have met a man who makes a business of following trout streams, fishing and hunting through the swamps, searching for frogs, and rare roots and herbs in their season. He finds ready market for Ginseng, American Ipecacuanha, Hellebore, or Indian Poke, from which is obtained a powerful cardiac depressent,—*Veratrum viride*, and species of *Cypripedium* also produce our native drug American valerian, which takes the place of the European drug, procured from *Valerian offinciallis*. Snakeroot, Dogwood, and various other plants afford excellent tonics. One can readily understand, as Thomas Wentworth Higginson remarks, "that many of our rarest flowers (in the vicinity of Boston) are being chased into the very recesses of the Green and White Mountains. The relics of the Indian tribes are supported by the Legislature at Martha's Vineyard, while these precursors of the Indians are dying unfriended away."[1]

Where years ago the swamps were fairly rose-purple with waving blossoms of the Grass Pink (*Limodorum tuberosum*) and Rose Pogonia or Snake-Mouth (*Pogonia ophioglossoides*), this year I found so few

The Pink Moccasin-Flower. (*Cypripedium acaule.*)

This is the only two-leaved *Cypripedium* found in the Atlantic region. It is closely allied with *Cypripedium guttatum* of Alaska and with *Cypripedium fasciculatum* of the Pacific slope. It is the most common species of this genus.

that I could readily count them. I discovered the possible secret of this extinction in the fact that a native of Etchowog was offered by some florist or gardener fifty cents a bulb or plant for all the specimens he could secure. This was an inducement for the vandal, but Nature cannot restore her species as fast as man can uproot them and devastate their haunts. Whether this is the true cause of extinction of these species in Pownal swamps I cannot ascertain beyond this inference; however, I am convinced that a small fortune has disappeared, estimated on the lost plants at fifty cents each.

Nearly all of the public schools are instructing the children in drawing,—teaching them to study the wild flowers as they find each in its season. Educators in all nature study urge the children to bring fresh specimens, and thus unconsciously encourage the extinction of the rare species of plant life in general. The children of each district school thus hunting over a limited area, soon, with childish strife, collect all the first and fairest flowers in their path. By the close study necessary, however, for the child to produce a drawing of the flower and its structural parts, a valuable lesson may in time be learned.

The story of fertilization, the necessity of the flower's producing seeds in order to continue its successive generations, will not be forgotten by the true nature student. But if the teacher were able to designate the rarer plants of her district, and teach her children the fatal results of continually gathering their flowers, she might awaken in the minds of the young people a higher reverence for the blossoms themselves, and scruples against depriving generations of children to come of their beauty.

There is hardly a child in the first grade in our schools who cannot tell the story of the bee and the Moccasin-Flower, and why the wonderful lines and dots of pink and gold are inside the downy shoe, instead of making the outside the more showy.

The first Moccasin-Flower which I found in Aurora's Bog in North Adams I gave to Ray, a little lad of my acquaintance, and he happily and proudly carried it to his teacher. When he came home, he could tell me that all these inner decorations of pink and gold were dewy-tipped with sweets, and were called "Honey Guides," just to invite bees within. And that although Master Bee goes through the front door of the Moccasin cottage, he somehow finds it locked when he wishes to escape, so in his excitement has to squeeze through the small back door next to the pollen-masses. He carries forth some of the pollen, and thus helps to fertilize the next blossom of this species, as he enters and rubs off the grains of pollen on the adhesive lobes of the viscid stigma. Insects thus are not permitted to rob the flowers of nectar and pollen without making a return for the food which the flower yields them.

Were it not for the bees and moths and various flies, the seeds of orchids would not mature, for it is a generally accepted fact that nearly all species of this family, wherever found growing, depend upon insect aid for fertilization and cross-fertilization. With the exception of one or two North American species of genus *Habenaria*, all other native species are aided by insects. These two species, *Habenaria hyperborea* and *Habenaria clavellata*, were, according to both Gray and Darwin, supposed regularly to fertilize themselves without aid of insects.

As the spikes of the Tall Green Orchis (*Habenaria hyperborea*) are frequent in the Pownal swamps, in company with the Showy Lady's Slipper, I became interested in this plant, so independent of Master Bee or Moth.

Professor Asa Gray, in various papers on fertilization of our native orchids, has said that they were all arranged for fertilization by the aid of insects, and that very few were capable of unaided self-fertilization. He tested several species, and proved that it might occur by accident, but in general his two self-fertilized species of *Habenaria* were still an unsolved problem, as later developments have proven in the case of his supposed self-fertilized species, *Habenaria hyperborea*, which he asserted "habitually fertilized itself." At least this species, although it may be fully equipped for self-fertilization, has been reported quite recently to be visited and fertilized by mosquitoes, proving that not in all instances is it found "habitually fertilizing" itself.[2]

In August, 1899, Professor C. A. Crandall, of the Agricultural College of Colorado, with a party of tourists camped on Medicine Bow Range, in that State, at an altitude of 10,200 feet, and observed abnormally developed mosquitoes bearing pollen-grains, which resembled those of *Habenaria hyperborea*; and so they proved to be, by subsequent experiments with specimens of this orchis gathered from a bog near by their camp.[3]

Another species of this genus, which is almost identical with the Tall Green Habenaria just mentioned, differs from it by bearing fragrant white flowers not adjusted for self-fertilization. This beautiful plant, *Habenaria dilatata*, grows sparingly in the choice haunts of the deeper Bogs of Etchowog, seeking frequently the pools near cold springs, and attracting numerous flies and moths by its rich perfumes, which one scents long before he discovers the flowers themselves.

Darwin mentions ten self-fertilized species of orchids for the whole world, and adds to that list ten more which were partially so, in case the proper insects failed to visit these plants in season.

He again asserts: "In my examination of orchids, hardly any fact has struck me so much as the endless diversities of the structure,—the prodigality of resources,—for gaining the very same end, namely, the fertilization of one flower by pollen from another plant. This fact is to a large extent intelligible on the principle of natural selection."[4]

Of the self-fertile species, Darwin remarks: "It deserves especial attention that the flowers of all self-fertile species still retain various structures which, it is impossible to doubt, are not adapted for

The Tall White Northern Orchis (*Habenaria dilatata*).

Near Arethusa's Spring, Bogs of Etchowog, Pownal, Vermont.

insuring cross-fertilization, though they are now rarely or never brought into play. We may therefore conclude that all these plants are descended from species or varieties which were formerly fertilized by insect aid."[5]

Darwin believed that, "bearing also in mind the larger number of species in many parts of the world which from this same cause are seldom impregnated, we are led to believe that the self-fertilized plants formerly depended on the visits of insects for their fertilization, and that, from such visits failing, they did not yield a sufficiency of seed and were verging towards extinction. Under these circumstances, it is probable that they were gradually modified, so as to become more or less completely self-fertile; for it would manifestly be more advantageous to a plant to produce self-fertilized seeds rather than none at all or extremely few seeds."[6]

Darwin questions: "Whether any species which is now never cross-fertilized will be able to resist the evil effects of long-continued self-fertilization, so as to survive for as long an average period as the other species of the same genera which are habitually cross-fertilized, cannot of course be told. ... It is indeed possible that these self-fertile species may revert in the course of time to what was undoubtedly their pristine condition, and in this case their various adaptations for cross-fertilization would be again brought into action."[7]

Indeed, the more this great scientist studied these strange flowers, the more he became impressed, and "with ever-increasing force, that the contrivances and beautiful adaptations slowly acquired through each part occasionally varying in a slight degree but in many ways, with the preservation of those variations which were beneficial to the organism under complex and ever-varying conditions of life, transcend in an incomparable manner the contrivances and adaptations which the most fertile imagination of man could invent."[8]

The extinction of species of orchids is due to causes inharmonious with Nature, therefore, more than to the failure of the insects in fertilization and cross-fertilization. Man and his bush-whack and bog-hoe are doing more toward the extinction of our rarer species of all plant-life in their continental range than any other natural element, in the swampy, mountainous districts of the East, as well as in the open swells on the prairies of the West.

The late Grant Allen expressed regret that the native Yellow Lady's Slipper of England, *Cypripedium calceolus*, "lingers in but two places," one of those stations being on "a single estate in Durham, where it is as carefully preserved by its owner as if it were pheasants or fallow-deer."

The wind, rains, and flowing streams, the birds, as well as migration and immigration of the nations over the world, are ever unconscious bearers of the seeds of our rare flowers and common dooryard weeds; yet for the rarer species Nature is indebted to the insects for the important process of cross-fertilization.

In country towns of New England, where summer resorts for tourists are numerous, one finds youthful venders selling the roots of the Orchid Family to "lovers of flowers," and thus even the lovers of Nature aid in the extinction of the treasures and wealth of her soil.

Species of *Cypripedium* are indeed the most gorgeous among our native orchids, and will be among the first of the family to become extinct, since they do not reproduce seedlings abundantly, even in their most choice haunts.

Notes:

[1] Thomas Wentworth Higginson, *The Procession of the Flowers*, p. 47.
[2] Gray, *Fertilization of Orchids*, in *Sill. Journ.* 1862-1863.
[3] C. A. Crandall, *Plant World*, p. 6. Jan., 1900.
[4] Darwin, *Fertilization of Orchids*, p. 284. 1895.
[5] *Ibid.*, p. 291.
[6] Darwin, *Fertilization of Orchids*, p. 292. 1895.
[7] *Ibid.*
[8] Darwin, *Fertilization of Orchids*, pp. 285-286. 1895.

The Showy Orchis. (*Orchis spectabilis.*)

The first orchid of the spring, found near the rocky borders of Thompson's Brook, East Pownal, Vermont.

5
The Queen of the Indian Moccasin-Flowers

The rounded world is fair to see,
Nine times folded in mystery;
Though baffled seers cannot impart
The secret of its laboring heart,
Throb thine with Nature's throbbing breast,
And all is dear from east to west.
—Emerson, *Nature*.

Between May 30th and June 8th, I made short excursions to the Bog of Oracles above the Glen of Comus. On the latter date I found my first blossoms of the season, of the Showy Queen of the Moccasin-Flowers (*Cypripedium reginæ*), the white sepals and petals standing fully unfurled, but still lacking the rich magenta-pink on the crest of the slippers which another week's time would give them. One feature this season, among these plants, was the unusual number of two buds on a single scape. While a single blossom is generally found on a stalk, I discovered now that nearly every other stem bore two buds.

At the same time and in the same place, along the edges of decaying logs on the borders of Ball Brook, grew the spikes of the Tall Green Orchis (*Habenaria hyperborea*). Its greenish-yellow color is conspicuously different from the tones of its distant relative, the showy, white-petaled queen of this swamp. Another spike similar to that of the Tall Green Orchis, but short and smaller in every way, stood near. It was not so tall and coarse as its sister species, and may have been a stray specimen of the Tall White Habenaria (*Habenaria dilatata*). These two species are peculiar in appearance, and many inexperienced bog-hunters would pass them by as weeds, and homely weeds at that.

Upon closer scrutiny, the peculiar twisted seed-pods of these flowers suggest a rarity. The name *Habenaria* signifies "a rein or thong," derived from the shape of the labellum in some species of this genus. They are often also called "Rein-Orchises."

On June 10th I drove into the Chalk Pond region, on the "Witch Hollow," or Gulf Road leading to the Centre-of-the-Town; and hitching old Bonny, took a circle around the peat and marl meadows, searching for signs of the Showy Orchis (*Orchis spectabilis*), a species of a sister genus of *Habenaria*. The Showy Orchis is due here about May 25th, the date on which the early Moccasin-Flowers awaken.

Four species of this genus unfold upon almost the same day. The Ram's-Head Cypripedium should bloom first, according to general reports of botanists, the Pink Acaule immediately follows, and the Larger Yellow Moccasins, and, at the same time, the Small Yellow Fragrant Slippers unlace their beautiful twisting petals. The Showy Orchis is supposed to be the first orchid of the spring to blossom in New England.

I discovered nothing in the Chalk Pond meadows, however, save that it was one of the most charming little corners in the town, showing deep erosions about its terraced basin, proving that the ice currents of the past flowed through these gulfs with terrible force.

I have found the Large Yellow Moccasin-Flower growing in close relationship with the dwarf fragrant species (*Cypripedium parviflorum*), in the Swamp of Oracles, in District Fourteen, about May 25th; while they appear later in the upland woods,—from June 6th until June 25th. They grow, as will be observed, along high, rocky hillsides as well as in damp, sphagnous marshes. The upland species are often found in open clearings on hillsides, among the dead brushwood heaps, where grow the Maiden-Hair and Christmas Ferns. Often they are in full sight, but sometimes they are hidden under small hazel-nut bushes, amid sapling white birches.

There seem to be three different forms of the Yellow Cypripediums, although there are but *two* accepted distinct North American species north of Mexico; these appear also to intergrade frequently. Close association of habitat has probably something to do with this cross-fertilization of the two species.

Finding the two marsh plants, *Cypripedium hirsutum* and *Cypripedium parviflorum*, growing side by side in the Swamp of Oracles, I observed a marked intergrading,—the larger species, *Cypripedium hirsutum*, producing variegated sepals and petals, or possibly now and then a brown-pink petal or sepal, imitating the type species of the smaller Moccasin-Flower. Both species were fragrant in a slight degree, *Cypripedium parviflorum* being, of course, the more fragrant of the two.

There is an European Yellow Cypripedium (*Cypripedium calceolus*) which is almost identical with the smaller species of North America, *Cypripedium parviflorum*. As early as 1760, *Cypripedium calceolus* was described and illustrated in color in Philip Miller's *Figures of Plants*. Linnaeus, 1740, gave the European yellow species the present generic and specific designation. Any history relating to that species of Lady's Slipper, as it was first known in Europe by Dodoens as early as 1616 under the title of *Calceolus Marianus*, will also pertain to the history of the two closely allied Yellow Cypripediums found in North America.

The common English name "Lady's Slipper" arose from the Latin *Marianus*, referring to "Our Lady," the Virgin Mary, while *Calceolus* is the Latin for shoe or slipper. Linnaeus, however, in 1740, being a devout Lutheran, objected to this species being dedicated to the Mother of Christ, and re-established the custom of dedicating the names of flowers to gods and goddesses of classical mythology known before Christ. The origin of the generic name *Cypripedium* is from the two Greek words Κυπρισ, an ancient name for Venus, and ποδιον, a sock, buskin, or slipper.

Venus, in classical literature, was also known as "Our Lady," the "Divine Mother" of the Romans, so that the common name has never in reality changed since 1616, when it was first applied to these shoe-shaped flowers of Europe, in honor of Mary, "Our Lady," the "Divine Mother" of all nations.

The Algonquin Indians, in their forests of Northeastern North America, saw this same shoe-shape resemblance in these flowers, and called them *Mawcahsun* or *Makkasin-Flowers*, since they reminded them of little Indian Moccasins. Thus arose the common name Indian Moccasin-Flowers for all our native species of *Cypripedium*. Lady's Slipper is distinctly of European origin, while Moccasin-Flower is most appropriately American, since this name was given by the first inhabitants of our shores, as it were, in mythological days. May the name of the Indian's Moccasin-Flower pass down through the coming centuries in honor of a race that will disappear long before these flowers, which they christened so appropriately.

I have never thus far found the Dwarf Fragrant Moccasin-Flower, an upland flower, which Higginson describes as growing on the "Rattlesnake Ledge" on "Tatessit Hill,"[1] in the neighborhood of Boston. The larger yellow species, *Cypripedium hirsutum*, grows in the Hoosac Valley high on the steep sides of the Domelet, while the smaller species seeks the deepest parts of the Swamp of Oracles and Aurora's Bog. I have collected it also in damp, marshy woods in Mosholu, near New York City.

The Large Yellow Moccasin-Flower seems, of the two yellow species, the more generally distributed over the continent, although most botanists state that the smaller species is the commoner. The dwarf yellow species is certainly the rarer plant in New England. In the Hoosac Valley, particularly in Pownal swamps, it is quite as rare as the Ram's-Head Cypripedium. I have discovered only one swamp here where it grows.

It will be of interest to make note of two species of our Eastern Cypripediums, which extend nearly to the Arctic Circle northward, as well as adjusting themselves southward near the Tropic of Cancer. One of these species is the Large Yellow Moccasin-Flower, reported as found associated with the Pink Acaule, in latitude 54° to 60° North, by Dr. John Richardson on Captain Franklin's journey to the Arctic lands in 1823. [2]

The Small Yellow Fragrant Moccasin-Flower. (*Cypripedium parviflorum.*)

The only really fragrant *Cypripedium* of the Atlantic region, closely allied with *Cypripedium Montanum*,—the Fragrant White Lady's Slipper of the Pacific slope. The plate shows the undulating sepals and petals as well as their rich brown-pink coloring. The two lower sepals are imperfectly united and are bifid at the apex. This species is almost identical with the European species *Cypripedium calceolus*,—the first *Cypripedium* described by Linnæus in 1740-1753.

Dr. F. Kurtz, in an Arctic Expedition in 1882, collected the large yellow species, *Cypripedium hirsutum*[3] of the Atlantic Region, as well as *Cypripedium passerinum*, which is endemic only to the Northern Pacific Region. *Cypripedium hirsutum* also extends from New England westward much farther than the pink species, *Cypripedium acaule*. The dwarf yellow, *Cypripedium parviflorum*, closely follows the larger yellow species both southward and westward, but according to the stations reported to the author for the continent, it cannot be said to have the broader range of the two species.

The Dwarf White Moccasin-Flower (*Cypripedium candidum*) may also be counted with Ram's-Head Cypripedium as one of the rare species of the Northern Atlantic Region. It is seldom found in the New England States. In the range reported to the author for this species, there is but one New England station. This has been given by A. W. Driggs of East Hartford, Connecticut.[4] This orchid belongs more especially to the damp swells of the prairie. It is very similar to the Dwarf Yellow Cypripedium, except in color, and like it produces a faint fragrance. This dainty white shoe is often no larger than the tiny Ram's-Head flower, the plant being about six to ten inches high, bearing small waxen shoes, the shape of the blossoms of *Cypripedium parviflorum*. I have often received descriptions from country lads, supposedly of these White Moccasin-Flowers, only to find that they were either *albinos*, or bleached out and pale specimens of the gorgeous colored *Cypripedium reginæ*. Often the latter seem pure white to the hurried observer in the swamps, for the albino or white variety rarely occurs. I found one plant, however, this season bearing two blossoms, the first I ever saw, and I removed the plant to watch it in my garden.

After Decoration Day, I had all I could do to keep pace with the unfolding flowers in the woods on Mount Œta. In the Chestnut Woods and Rattlesnake Swamp region, near Lloyd Spring, and along the mountain sides of the Knubble and Domelet, I found beautiful azalea shrubs laden with luxuriant clusters of fragrant pink flowers. These open woodlands become brilliant with these rose-colored blossoms. The Large Yellow Moccasin-Flower was here too, with violets, Stars-of-Bethlehem, and innumerable pink blossoms of *Cypripedium acaule* growing along the side hill, shining out from every corner. All at once, these nearer woodlands had unfurled their banners of spring, and now, "With blossom, and birds, and wild bees' hum," they held me from the more distant Bogs of Etchowog. On the 14th of June, however, I decided to take old Bonny and the buggy, and drive to these bogs to see if any Pogonias and Limodorums were budded as yet amid the grasses of the open cranberry marsh.

Bonny hitched to the old buggy, my faithful old Major at my side, and I, with my vasculum for rare flowers, a basket containing drinking glass, carving knife, and bog-hoe for gathering special roots, started down the hill on an easy trot toward Pownal Pond. As I passed School Fourteen, I was cheered and hailed by the children, who shouted,

The Small White Moccasin-Flower. (*Cypripedium candidum.*)

This species is especially an orchid of the damp swells of the prairie, growing in company with the Painted Cup and Iris.

There, I think, on that lonely grave,
Violets spring in the soft May shower;
There, in the summer breezes, wave
Crimson phlox and moccasin flower.

Bryant

"Going a-flowering?" I nodded "Yes," with a "Get-ty up" to old Bonny, who had thought I wished to visit along the way.

It was warm and dusty, and whenever I could, I drove through the streams which crossed the road, in order to swell the felly, and thus tighten the tires to my rattling wheels. Although I felt that by driving along the highway I was losing much beauty that was unfolding in the fields and fence corners, I found this method of progress quite comfortable.

How these East Pownal bogs came by the musical name of *Etchowog*, I am not quite certain; nor do I know exactly what it means. It may have come from a primitive language of a mythological age for all I know, or it may have come from the Itch-Weed or Indian Poke and Poison Rhus, which cause much irritation of the skin. I am safe in saying that it is a corruption of the Indian's Greek and Latin words for "itch" and "bog,"—at least this etymology quite suits the designation of these swamps. Ever since I can remember I have heard the older folk of the town call it Etchowog. I have associated the region with rare flowers, orchids, pollywogs, snapping-turtles and mud-holes, together with the schoolhouse in District Thirteen, where the good people hold Advent meetings, and set the dates for the world to come to an end. To me it seems one of the brightest, richest of swamps, full of "Bottomless Dead Holes," where only bull-frogs peep and trill and croak the whole season through, till their notes blend with the chirp and whirr of the autumn crickets.

At the Barber Mill, I hitched Bonny to a fence-post and started on my excursions. I looked through the open meadow east of the mill to see if I could find any rose-colored Pogonias and Grass-Pinks. There was as yet no sign of them; so I came back to the mill and turned in through the bars, on the north side of the pond, where I followed a grassy path around the hill to the treacherous Cranberry Swamp farther northward, where I had been cautioned not to wander alone.

Sounding the margin of the marshy meadow, I found quaking and unstable ground. With a ten-foot pole I probed the depths of the mud, and found it unfathomable, and no signs of *terra firma* about it. Pickerel-weed, eel-grass, frog's-bit, and the leaves of arrow-head grew about the pools. I could not very well find an entrance here, unless for a permanent residence. So going northward along the west shore of this mud-pond, I came to a place which promised fair and safe walking, with my waterproof boots for protection. At first I felt my way very cautiously, then grew bolder and forgot that I was in a dangerous place, for the farther I advanced, the firmer and drier and more enchanting became the field of my vision.

Before me opened a wide expanse of meadow-land, where even unruly cows dared not wander, and man seldom ventured to trespass. Nature's remote solitude indeed was peacefully hidden here. No human voices nor sounds of hay-making ever echoed over these luxuriant fields, and the grasses grew sweetly, to fall untouched to earth again, mown as it were by the autumn winds, and stored beneath the drifts of November snow, to lay, in time, one more thin coat of soil upon the unplumbed depths of this ancient lake bed. During some long-ago winter, some one had ventured here while the earth was frozen and safe, and had built a homely hedge-fence through the meadow, probably to keep the cattle pasturing hereabout away from the dangerous bog. This fence was the only visible trace of man. In its tumbled-down and overgrown condition, it became a part of Nature's self, and added to the picturesqueness of the field. Although Rafinesque says "that he hates the sight of fences like the Indians," to me the hedge-fence is one of the wildest and most primitive of forest barriers. Indeed, it must have originated with the veritable wild man himself.

I was tempted on and still farther on through the meadow, by the brilliant crimson-purple blossoms of the Pitcher Plant, or Side-Saddle Flowers, so named on account of the hard shells of the stigma of these flowers resembling the padded cushions of a lady's ancient side-saddle. This cushion was known as the "pillion." The more common name in this locality for these flowers is St. Jacob's-Dippers and Dumb-Watches, children playing with the hard shells of the stigmas left after the purple petals have fallen, calling them watches. The convex surface of the stigma does indeed resemble the face of a watch, although there are no hands to point the hour. Gay blossoms of Fleur-de-lis flaunted their gaudy petals, and many times deceived me by making me imagine that I spied the Purple-Fringed Orchises in the distance, waving amid the tall grasses.

Here I dreamed away an hour or more, following out some little paths, worn perhaps by the muskrats or swamp minks or wicked weasels, or perchance by the tiny feet of the meadow-moles, who apparently had blindly rooted various underground tunnels in every direction. I can fancy them all trotting swiftly along, playful at times, yet with an eye to their affairs,—quite as important in the scheme of Nature and Science as are the brokers' studied operations in Wall Street. The weasels and minks are the terrors of the other path-holders in this natural syndicate. They are indeed the high and dreaded trust officials of the lesser and blind rooters of the earth.

Tangled vines of the marsh cranberry were now in full bloom, and at the same time the soft fruit of last autumn's crop was present on the vines, still bright crimson, even after enduring the winter's frosts and stubborn snows.

Looking northward to see what fields lay unexplored beyond me, I realized the remoteness of this region slumbering amid these glacial hills. To my right towered the Dome, the highest mountain of Pownal, of a bluish-green tone, against the sky. Nearer, graceful elms, tall pines, and numerous low pointed, lighter green tamarack trees lifted their spires, and adorned the distant meadow; while in the wide expanse on the west side, along the edges of the swamp, rose the giant forms

The Queen of the Indian Moccasin-Flowers (*Cypripedium reginæ*.)

From the Bogs of Etchowog, Pownal, Vermont.

of elm and pine, and tall, lithe trees of the swamp maple, flashing forth their crimson and gold blossoms, reminding me of the coloring of autumn leaves. The nearer marsh was rich with tasselled grasses and blossoming vines, dotted here and there with the cardinal buds of the Pitcher Plant and the purple Fleur-de-lis. It seemed a land of dreams.

The air vibrated with the happy, mellow song of birds, interspersed with the ever-present lesser sounds of deep solitudes. Major, like me, at first, was cautious where he wandered, but once amid the various haunts of wild creatures of the wood, he caught the happy spirit of the hound, frisking and studiously following the paths of the wild little animals to the very doors of their homes.

To test the land, I stood and deliberately shook the foundation of the earth. All the blossoming ground about me, for at least fifteen feet distant, trembled as if it were so much jelly. Yet the spot was honey-combed and dry on the surface, there having been little rain in this region during the month.

I now sought the western hillside path, and bearing northwestward around the border of the swamp, I occasionally ventured in and out along the edges of the meadow bushes. Finally I reached the swamp maples, which I had observed from the interior, and I secured a good-sized branch of the gold and crimson clusters to carry off with my load of treasures. On every hand, out of the small, muddy pools of water, rose the leaves of the Buckbean (*Menyanthes trifoliata*). The beautiful spikes of white-bearded flowers were turning brown with age, and the plants were setting their bullet-like seed-pods. Now and then, beneath the low, shaggy pines, I found the humble Pink Moccasin-Flower (*Cypripedium acaule*), which I hailed as a sign that the Showy Queen of the genus might dwell not far distant.

Knowing the favorite haunts which this orchid seeks, I searched through all the dark corners of the swamp. At the extreme northwestern portion of the region, I entered a dense shaded corner about fifty feet square, where were many springs soaking through the sphagnum to the deeper fields of the interior which I had so lately left. Here were numerous decaying pine and tamarack logs, low sapling willows tangled amid the small scrubby spruces and tender pines, which were striving against the greater natives of the forest to lift their spires as high as possible; but however eager they were, they had not attained a height above ten or fifteen feet at most. Many were already discouraged or had died in the competition, and their wasting forms were still standing with broken and weather-worn trunks and limbs.

Tall brakes and Indian Poke ran riot among the deeper mounds of moss, which covered the decaying roots of the long wasted primeval pitch pines. The dark, sluggish pools reflected weirdly the ferns and trees above them.

Shooting up from these piles of sphagnum, I found at least fifty plants of the Showy Moccasin-Flower (*Cypripedium reginæ*). They were pregnant with slumbering buds, and would surely be in full blossom by June 20th. Happy over my good fortune at locating another station for this species, I prepared to bend my footsteps toward my horse and buggy,—glad indeed to know that I would not be obliged to walk home, laden as I was with Pitcher Plant roots and various other shrubs and vines.

Near the mill, just north of the little bay in the pond, I found quantities of the Yellow Pond Lily or Spatter-Dock (*Nymphæa advena*) just beyond my reach. Securing a long willow sapling with a tender end, I tied it into a loop, and stepping out into the shallow edges of the pond to an old pine log, I snared off several of these golden cups, which the children call Cow-Lilies. I floated them in to the shore, where I soon gathered them up and packed them in my vasculum.

A glance into the water along the edges of the old log revealed thousands of tiny pollywogs or tadpoles, as well as half-formed frogs, the hind legs beginning to put forth on the large tadpoles. Here, basking in the sunshine, were lizards, snails, leeches; and various species of small fish were sporting in the shallow waters. Perch, suckers, and eels are plentiful in Pownal Pond, which is locally called Perch Pond, from the abundance of perch found in its waters. These fish seemed to seek this sheltered arm of the pond to leave their young fry under the sheltering lily-pads.

Near the projecting stumps, amid floating logs were snails' eggs, and I noticed several baby turtles, recently hatched from eggs in the sand, varying from the size of nickels to that of a silver dollar. Eelgrass and many marsh grasses and sedges grew or floated on the water, among which the small fish could hide.

On the edge of the water among the ferns and brakes I found the leaves of the Purple-Fringed Orchis (*Habenaria psycodes*), but no plants likely to bloom this season.

When I reached the mill, I placed my treasures in the buggy, and started after that part of my load which I had left around the hill. On my return, I gathered some waxen, crimson cones of the beautiful tamarack tree by the path. When I bade farewell to little Merwin and his mother, who lived in the mill-house, I asked them to watch for the rose-purple orchids,—Pogonias and Limodorums,—which were now due any day, east of the mill. The boy was very earnest and observing, and I knew that I now had a comrade to guard over the Bogs of Etchowog.

Students from Williams College, and tourists from near and afar seek these swamps of Pownal for botanical specimens, and Merwin had often been their guide to the haunts of these rare treasures. He told me that students from Williams had, the year before, gathered innumerable pink and purple flowers in these marshes, as well as the beautiful bearded spikes of the Buckbean.

For a succession of years—during all of President Carter's term at Williams College at least—it has been the unique custom to bank the chancel of the Congregational Church with the Showy Moccasin-Flowers and Maiden-Hair Ferns, on Baccalaureate Sunday,—which occurs usually about June twentieth. These gorgeously colored orchids reach the height of their perfection about this date. They seem a fitting decoration for the church during the Commencement services of this college, situated in the heart of these Hoosac Highlands.

Plentiful as are the colonies of this Showy Moccasin-Flower in its pet localities, it has always been an interesting question to me where the great numbers of perfect blossoms grouped about the chancel are secured. They are known to the children in each school district, and usually they are collected as soon as discovered.

It is surprising to me that extinction of this rare plant is not taking place more rapidly hereabout. This orchid produces very few seedlings in its native haunts, and at the rate of collecting both its blossoms and roots in this valley, we must surely look for total extinction in less than half a century more, unless this ruthless plucking is modified.

Notes:

[1] Thomas Wentworth Higginson, *The Procession of the Flowers*, p. 17.

[2] John Richardson, M.D., *Bot. Appendix, Report of Franklin's Journey*, 2d ed., p. 34, 1823.

[3] Dr. F. Kurtz, *List of Alaskan Orchids*, Expedition 1882.

[4] A. W. Driggs, *Catalogue Plants of Connecticut*, p. 19. 1901.

6
Hail-Storms at Etchowog

. . . Suddenly, a flaw
Of chill wind menaced; then a strong blast beat
Down the long valley's murmuring pines and awoke
The noon-dreams of the sleeping lake, and broke
Its smooth steel mirror at the mountain's feet.
—Whittier, *Storm on Lake Asquam.*

On June 21st, with Major I walked down through the Swamp of Oracles in District Fourteen, along Ball Brook to the Kimball Farm bogs, and so on once more to the Bogs of Etchowog and the new colony of *Reginæ*—the queen of the Indian Moccasin-Flowers—which I had so recently discovered in Cranberry Bog north of the pond. I found prime blossoms all along the tiny path, in the course of the stream through the deeper parts of Glen of Comus, and in the Kimball Bogs, and I was in hopes of finding them in the swamps of Etchowog.

As I passed through the sphagnous meadows east of Kimball's barns, around the hillside path to Arethusa's Fountain, I noticed several flowers of the Cypripedium I was seeking, and recognized the leaves and green-budded spikes of *Habenaria psycodes*, which would later, when fully in bloom, change to a delicate purple.

I made use of the fence boards to walk through the muddy portions of my path. I had learned by former experiences here to avoid the "dead holes." Stepping on some boards just above a muddy pool, and suddenly turning, I was happily surprised to see many spikes of the Tall White Northern-Orchis (*Habenaria dilatata*) standing near. The air was full of their rich perfume, and many small flies and moths hovered around them, sipping the nectar. I gathered a few spikes, and went on to the cool spring beyond, finding meanwhile an abundance of wild strawberries along the borders of my path. These were very large from growing in the moist shade.

On the hillside, up which I climbed to the west for a short distance, I found pretty leaves of grasses, delicate emerald in color, growing in a triangular form, and resembling lily leaves.

The Small Purple-Fringed Orchis. (*Habenaria psycodes.*)

I had heard distant thunder rolling off to the northwest, and it caused me to hasten onward. My rest, therefore, at the spring was brief to-day; although so far away from home, I was not so far from shelter, and the thought of a shower was welcome, for the air was sultry. As I neared the open swamp, beyond the mill, the storm made rapid strides, but I wandered up and down the meadow long enough to assure myself that this season the Pogonias and Limodorums were not in bloom on time.

Large drops of rain began to fall from the black clouds, and as I hurried toward the shelter of the mill, I met Merwin and his mother returning to their home. They motioned me to join them. As I did so, great gusts of wind dashed over us, and suddenly huge hailstones pelted the earth. Leaves and small twigs and young apples fell on every side, while the half-grown nuts from the Butternut-tree (*Juglans cinerea*), in the dooryard, were soon stripped away, with the leaves and broken limbs of the tree. Some of the hailstones were the size of small hen's-eggs, perfect, oval ices which might have been turned out of glass moulds.

Soon the air became very chilly, as during the first snow on a damp November day, while the ground was white with hailstones. This abrupt change in the atmosphere from heat to extreme cold caused untimely deaths in the chicken yard. The old mother hen lost her head completely, and unable to find shelter in the barn because of the banging doors, she put her head in a crevice while her brood ran about and perished with cold or were killed by the stones.

Merwin's mother sadly watched the devastation of her little garden, and the death of her chickens. It was impossible to go to their rescue without danger to our own heads. This storm continued about two hours, alternating now and then with a calm, to return again and again with sudden fury. At the end of that time, although it still rained sadly, I started for home, knowing that with rubber boots I could wade, if necessary, through any ordinary streams.

The weather had turned so cold that an icy coating covered the meadow grass and the borders of the road, and promised not to melt away in haste.

As I neared Kimball Farm, where Ball Brook meets Thompson's stream, I found the road opposite the barns flooded,—like a river flowing across the road. It was far too deep for me to wade through, besides, the current was so strong that I should have been tripped had I ventured it. I had to walk some distance on the stone wall and over a heavy plank, which some one during a previous deluge had placed here for a high-water footbridge in an emergency.

A walk up the hill, and I turned off the road, entering a path through the cow-pastures, to see the heaps of hail under the pines along Thompson's Brook, which was a beautiful, roaring and seething torrent now, as it plunged and leaped down through its rocky flume to the valley below.

As I came out on the highway again, at the bend in the road near Ball Farm, I heard the familiar voice of some one who had been sent in search of me. I was warmed with enthusiasm and interest in the storm's ravages, and thoroughly enjoying my walk. However, I was grateful for a ride home. Passing by School Fourteen, we saw the prudent teacher scanning the sky before she ventured forth. We noticed many broken panes of glass in the schoolhouse windows, while dozens were shattered in the houses along the way.

I had hoped to revisit the colony of the Showy Moccasin-Flowers which I had found in Cranberry Swamp, north of the pond on June 14th. But Merwin's mother told me that without doubt they had been gathered on Saturday afternoon, June 19th, by three students from Williams College; she had seen two of them come around the hill by the pond about five o'clock on that day, bearing a new bushel-basket filled with these gorgeous orchids, while the third soon followed laden with more than he could easily carry far in his arms. They followed the cool mountain road over the Domelet to Williamstown, a road over which the yeomen from northern Berkshire were led to battle at Bennington, on the 16th of August, 1777. The road is seldom traversed now, and at best is rough and rocky. It leads directly from Bennington southward to North Adams, under the mountains, and indirectly to Boston.

Had the storm come on Saturday, instead of Monday, very few blossoms of these orchids would have decorated the church chancel on Baccalaureate Sunday for Williams' Commencement exercises.

The fact that these students come to the Pownal bogs for these orchids assured me of the scarcity and rarity of the species in Williamstown, although they may be found sparingly in the swamps of The Forks along Broad Brook, just over the Vermont State Line in Pownal. This stream rises on the east side of the Majestic Dome, and flows down to the Hoosac by way of White Oaks, and thus enters Williamstown, where it soon joins the river. The orchids in The Forks are quickly plundered, long before June 20th, by ignorant tourists or students afield botanizing, who either do not realize or do not care that plucking all these rare blossoms will in time bring about their total extinction.

Orchids may in many instances produce seeds in abundance, but why they do not reproduce more seedlings is a problem not easily solved nor remedied.

Darwin once estimated that a single spike of the English Orchis (*Orchis masculata*) produced over 186,000 seeds, and that at this rate its grandchildren would soon carpet the earth; while Müller says also that his brother estimated 1,750,000 seeds in a single capsule of another species of the family (*Maxillaria*). We must remember that the species of *Orchidacea* are not as a rule self-fertilized, as are the more abundant and common flowers and weeds, which often cover acres of swampy land and fields of waste land. Our native orchids are wholly dependent upon insects for fertilization and cross-fertilization; yet, for some cause or other, comparatively few of the ripened and fertile seeds germinate and reproduce new seedlings. Our Moccasin-Flowers do not appear to multiply in many swamps, while species of Orchis and Habenaria are never abundant in this region.

For years now, I have noticed large groups of the Showy Lady's Slippers growing in Rattlesnake Swamp near Lloyd Spring, and I can find little increase in the number of plants, or the size of the old snarl of roots. In fact, they seem to be diminishing in numbers.

There is an old colony in this region that has stood for about seventy-five years, much the same in size, on the authority of the old inhabitants of this, neighborhood. It stands to-day among the shrub-like willows and swamp maples, at the feet of little scrub pines and dwarf double spruces, hidden from the sight of travellers in the path by a prostrate tree trunk and decaying primeval pine stump. I observed this colony years ago, and this season it appeared the same to me, occupying a space about two feet square. I counted forty-two full-grown flowers, many stems bearing two blossoms. This indeed was one of the most charming sights, suggesting the luxuriance of the humid climate of the tropics. It was even more enchanting than the colony of Pink Moccasin-Flowers,—that famous group of two hundred buds which the children in District Fourteen secured ahead of me, since this group of flowers were massed more closely together. I wished a sight of the Pink Moccasin-Flowers at their best. I left these, too, undisturbed save by the little moths and mosquitoes and honeybees, which came to drink the nectar within the pearly pink and white cups.

Notwithstanding the recent hailstorms, which had split many cups and spilt the dew, the flowers were developing plump, hard seed-capsules. Thousands of fertile seeds must fall and fly about from this colony; and yet the aged snarl of roots remains the same.

A unique row of seedlings of this species (*Cypripedium reginæ*) too young to blossom, and reminding one of a row of barn-swallows, not yet sufficiently matured to fly, grew along a moss-covered pine log, near the parent colony of plants. Digging down, I found the old log about twelve inches below the surface. It was sound at the heart, bare of its outer bark, and had become so imbedded in the water-soaked peat as to be absolutely preserved. The stump from which this tree had fallen was worn and crumbled away to the very earth, and capped with moss. It will require years for this log to settle into the peat deeply enough to allow these seedling orchids to ply and mass their roots in generous soft soil. Unless their roots deeply penetrate rich soil, the plants become pale in color and dwarfed, like the plants growing in loose sphagnum.

I missed some old colonies; these were of a new generation, and if they are not starved out, will blossom here in a row another year.

Another cluster of plants growing near by produces the deepest magenta blossoms that I ever beheld, and only in this one group have

The Showy Moccasin-Flower—The White-Petaled Lady's Slipper—
The Queen of the Indian's Moccasin-Flowers. (*Cypripedium reginæ.*)

Rushes tilting their burnished spears,
These are her courtly cavaliers.
Heart of my heart, we forswear the rose;
We have been where the lady slipper grows.

Clinton Scollard, *In the Heart of June.*

I seen this particular hue. A deep rose-purple extends over almost the whole labellum, and from a distance I thought I had discovered the long-sought Purple-Fringed Orchis,—such a flame of color rose before me. It almost seemed a variety of the true *Cypripedium reginæ*.

This swamp produced just one hundred blossoms this season. Of this number I gathered about twenty-five among the scattered plants, leaving the older groups to ripen their seeds, if possible.

I found the first fully unfolded Showy Lady's Slippers of the season, on June 8th, in the Swamp of Oracles in District Fourteen; while those of Rattlesnake Swamp unfolded fully this season on June 20th, and faded about July 1st, the season being shortened by the heavy hailstorms.

I have noticed that orchids growing in open, sunny swamps are stocky and short-scaped, bearing highly colored blossoms; while in shaded, muddy glooms the plants are rank and tender, with pale flowers, which do not last nearly so long as those which grow in the sunlight. The deeply colored specimens mentioned above grew wholly in the sunshine, and beside a fresh flowing stream.

I have transplanted all the New England species of Cypripedium, but only two of them took kindly to the garden for a succession of seasons. The small yellow species, *Cypripedium parviflorum*, seems easily naturalized in our damp woodland corners of the garden. The large yellow species, *Cypripedium hirsutum*, closely allied with the small yellow species, is easily managed in the same colony. The Ram's-Head (*Cypripedium arietinum*) is more choice in its home, being rarely seen in cultivation. It is not very plentiful even in its native haunts.

I have sent plants of the Showy Lady's Slipper and the Large Yellow Lady's Slipper found on Mount Œta, to New Bedford, Massachusetts, to Herkimer, New York, and to New Haven, Connecticut. In every instance they have become happy in their new surroundings, thriving and blooming through several seasons. The Small Yellow Cypripedium in New Haven has flourished and bloomed for ten seasons. The seed-capsules of these orchids, however, have never matured fertile seeds in this garden; and the pods wither up and do not develop as in the forest bogs, for want of the proper insects to fertilize them. It would be well to secure pollen from sister species of this plant in the Swamp of Oracles, and insure fertilization and cross-fertilization of this tame garden plant. We might look for possible hybrids, since this species is well broken away, by ten years of cultivation, from its primeval condition.

The Showy Lady's Slipper does not take so naturally to the garden, and in many instances does not live so long in captivity as would be expected. It will, however, produce seedlings readily, if care is taken to protect the surrounding soil in winter, where the seed is sown.

An interesting experiment, with artificial agencies producing fertile seed of this species, is related by F. F. Le Moyne of Chicago. He sowed the seed thus obtained artificially for two successive seasons, and secured seedlings from each sowing. He also believes that "this plant could be multiplied very rapidly from seed thus fertilized," in garden culture.[1]

This year, I sent the rare Ram's-Head to the New Haven Garden, with hopes of its blossoming next May. This Cypripedium is the rarest orchid in North America.

The Pink Moccasin-Flower (*Cypripedium acaule*) is the most common species of the genus in New England, and on the continent of North America, north of Mexico, with the exception of the two Yellow Cypripediums, which claim a broader range from east to west. The Pink Cypripedium proves the most stubborn and difficult in cultivation. It may be potted during the winter, but seldom, if ever, blooms more than a single season.

While many of our native orchids have a certain amount of adaptiveness to environment, they never will be found to choose absolutely dry soil, such as the rocky sheep pastures in which the common pennyroyal thrives. A sheltered, damp corner is safest for the exiled plant, where the sunshine searches long to brighten its petals.

One cold day in early March, I secured a frozen sod containing the roots of the Showy Lady's Slipper, and made an artificial bog in the bay-window, where I watched it thaw out. The flowers burst forth about a month earlier than when in the swamps. But although they were fully in the warm rays of the May sun, the blossoms were pale and delicate. The same cluster of plants sent forth deep rose-tinged blossoms the next season, in the damp corner of my garden, where I planted them. They became strong, healthy plants, flowering several seasons on the regular date for Pownal, June 20th. It is therefore evident that dates for blossoming differ more according to the exposure of the haunt than to the variations of seasons. But in the Swamp of Oracles I know where I can find this Showy Queen of the Indian Moccasins as early as June 8th, and I know of other haunts where it is not unfurled until the 15th and 20th of the month.

Notes:

[1] F. F. Le Moyne, *Garden and Forest*, 3: 1890.

7
Sweet Pogonias and Limodorums

Come bring me wild pinks from the valleys,
  Ablaze with the fire o' the sun—
No poor little pitiful lilies
  That speak of a life that is done!
Alice Cary, *Be Still.*

On June 26th we drove over to Thompson's Trout Pond. We took the old flat-bottomed boat, and with one slab board for a paddle, steered slowly over the whole surface of the lake,—a beautiful, clear little mountain mirror, with good-sized fish swimming about. I searched along the shores for the long-desired Purple-Fringed Orchises, but still without success. Fleur-de-lis grew abundantly about the lake; and in the little dents and bays among the sedges and cat-tails, I found the Yellow Spatter-Dock or Cow-Lily (*Nymphæa*), so named in the time of Christ by the ancient herbalist, Dioscorides, who first gave it the Greek name *Blephara*, and later, in Latin, *Nymphæa lutea* and *Nenuphar citrinum.* It was known in England in 1500 as Yellow Nenuphar or, Water Lily.

The swamp birds are tame and saucy here. Paddling our boat into the reedy shores among the alder bushes, where they were nesting, they seemed to take no alarm at our approach, but stood their ground pouring forth beautiful liquid notes. In one place near the centre of the lake, we crossed an expanse of deep water where long rootlets of the Water Persicaria (*Polygonum amphibium*) supported glossy carmine, lance-like leaves, which swayed gracefully on the surface of the swelling waves as we approached. These strange deep-water weeds send forth rich crimson or pinkish flowers a little later, seeming fairly to stain the lake. I had never seen this species before growing in such depths of water. It is a species of the Buckwheat Family, and a near cousin of the barnyard smart-weed and the knot-grass or door-weed. The generic name, *Polygonum*, comes from the Greek, meaning "many knees." It is so called on account of the swollen joints of some of the species of this family. The leaves of the Water Persicaria are brilliant crimson on the lower surface, and with age and exposure the upper surface turns deep Indian-red.

These plants were rooted at least fifteen feet below the water's surface in the mud. They may be found, too, along the shallow shores of Pownal Pond. They also grow in ponds and lakes far northward to Quebec and Alaska, and as far south as New Jersey and Kentucky, and westward to California. They thrive at an altitude of two thousand feet, in the lakes of the Adirondacks, blooming there, as a rule, in July and August. Thoreau observed this species in the lakes of the Maine woods, during his journey in 1853.

On the 30th of June I ventured forth to Etchowog, in search of Pogonias and Limodorums, although the season was almost too far advanced for prime specimens. I had heard the day before that some blossoms of these plants had been gathered in the Westville Swamps, near New Haven, Connecticut. I thus felt encouraged to search once more for these beautiful orchids. With luncheon and vasculum, and Major following me, I journeyed over the meadows and hills of Mount Œta to the north slope of the Domelet, where I crossed the country road. Finally I descended into a deep basin under the Dome, which rises east of the Domelet. Northward nestled the neat white and red farm buildings near Thompson's Pond, and far beyond them all I saw the blue, blue hills of Bennington County.

Everywhere I searched for the Fringed Orchis, which had so far eluded me in these swamps. The meadow seemed interminable as I circled around to the east of the pond. Bearing to the northward, I noticed nothing new except the ravages of the recent hailstorm. It had cut down flowers and corn-fields alike. The very hills were washed down from the mountain sides; great gutters and still flowing streams were eroding the corn-fields, scattering the sandy soil broadcast over the once green meadows. Even the edges of the grasses were brown and sear, and the Timothy-heads of the Cat's-tail Grass were stripped prematurely of their seed.

I followed Thompson's Brook, leading northerly from the pond, in through several willow and alder swamps. Then, instead of following down the rocky channel to Ball Brook Forks, I struck out directly at the head of the Meyers Road, over the fields, north from the maple-sugar house, and landed on the high hills south of the great meadows of Etchowog. Sleeping at my feet lay those sphagnous bogs which had already yielded me so many rare flowers, and so much pleasure. Northward stretched out a vast sweep of hills and valleys, reaching nearly the whole length of Bennington County. To the right towered the massive abutments of the Dome, and to the left rose the isolated form of Mount Anthony,—these two mountains framing, as it were, the gap northward, through whose wide vista I could define the dim blue heights of Mount Equinox, at Manchester. Nearer, I could trace fertile vales and sloping hillsides, dotted here and there with woodlands, scattered trees and farm buildings.

The Northern Gap. Showing the Taconic Mountains of Bennington County, from Mount Œta, Vermont. The Bennington Battle Monument towers to the left in the distance.

Standing still nearer in the shadow of Mount Anthony was Bennington Hill, with the Battle Monument clearly outlined even at this distance, some ten miles away. In the nearer landscape were discernible the serpentine windings of Ball Brook, with its long chain of tamarack and balsam-fir swamps, spreading out here and there toward Bennington,—where, I dare say, are many rich and undiscovered colonies of Lady's Slippers.

Nearer yet, the knob-like glacial hills around Pownal Pond shield the Cranberry Swamp to the north, and the open Bogs of Etchowog east of the pond. Nestling among the trees by the mill, I picked out the roof of the mill-house where little Merwin lives. But the shadows of hill and mountain were growing longer in the valley as the sun sank toward the west, and it behooved me to waste no more time dreaming on the hilltop. So I slowly descended to the valley, groping my way between bushy young pines, passing a herd of gentle, meek-faced Jersey cows feeding on the hillside. I found many cow-paths running around the bog, and was led out into the swamp at a point nearly opposite the little white schoolhouse of Barber District, Number Thirteen.

I did not find the place rose-purple with the little orchids, as it should have been, but I did find a few dozen plants of Grass-Pinks (*Limodorum tuberosum*), and six or eight delicate rose-pink blossoms of Snake-Mouth (*Pogonia ophioglossoides*). I gathered a few flowers of each, grateful that any remained to assure me that they were not quite extinct here, and I observed how very careful one must be in plucking the flowers not to pull the little roots and bulbs out of the moss at the same time.

All my plants grew east of the stream that runs through the centre of the swamp. When I tried to cross this creek, I found it so broad and deep and muddy that I could not get anywhere near it. Wandering toward the road skirting the bog, I came to a rude board bridge over the stream, indicating a path formerly leading through the swamp to Barber's Mill. Some high-water tide had twisted and turned the plank about so that only by catching and clinging to small bushes and saplings on the other bank could I succeed in crossing. I found no Pogonias and Limodorums on the west side of the stream, and it was just here that I had once found the meadow one wave of rose-purple.

Reaching the mill, I hastened around the bend in the road. A little to the south of Arethusa's Spring, and scarcely five feet to the left of the path, under some willows, I saw a dark, insignificant looking pool. Stooping down and touching the surface, I found it icy cold. This pool, Merwin's mother tells me, has always been here, and at no time in her memory has she heard of any one being successful in measuring its depth, although it has been probed with very long sounding-poles. These have been dropped fifty feet or more. Frequently she has left a long pole standing in the pool, only to find upon returning later that it had disappeared in the depths below, proving great suction. Such

The Rose Pogonia. (*Pogonia ophioglossoides.*)

A delicate little orchid, found as comrade with the Grass Pink, and frequently with Arethusa, in wild sphagnous meadows.

holes and springs are characteristic of the swamps of Etchowog, where the original lake bed was located over a century ago, before the water of Ball Brook was turned in its course through the present pond west of the mill. This "dead hole" should be fenced in and marked "dangerous," since it might so easily be stepped into by one unacquainted with its character.

I followed the familiar and loved path out to the sphagnous meadows east of Kimball's barns. Taking a straight line southward up the hill, back of an orchard, along the border of a field of Indian corn, I came again to Thompson's Brook, on its way to join Ball Brook, near the Kimball barns below. It is one of the stoniest channels, narrow and deeply worn, with here and there graceful clinging ferns slightly caught to the banks, and often completely hiding the huge boulders and ledges. Pines and hemlocks are the principal trees along this stream. The twisted and uncovered reddish roots of the hemlocks seemed to have split the black shelving slate rocks asunder with their growth. I threaded my way as near the brook as possible, often finding it necessary to wade in the stream until I reached the bend in the road near Meyers's sugar-kitchen among the maples. Here, turning to my right, I followed the shaded road leading past the schoolhouse in District Fourteen, and homeward to Mount Œta.

My orchids were pretty well withered on reaching home, and not in good condition for studying. These delicate species of Pogonia and Limodorum are easily wilted, losing their beauty and elasticity soon after being severed from their roots. These two species, Adder's-Mouth Pogonia and *Limodorum tuberosum*, are almost invariably found together,—comrades of different genera that travel far and wide in company throughout their continental ranges.

The Adder's-Mouth Pogonia has been formerly confused with our native species of *Arethusa bulbosa*, and for some time was known as Adder's-Tongue Arethusa. Thomas Wentworth Higginson writes: "On peat-meadows the Adder's-Tongue Arethusa (now called *Pogonia*) flowers profusely, with a faint, delicious perfume,—and its more elegant cousin, the *Calopogon*, (now called *Limodorum*) by its side."[1]

Yet Thoreau had a different impression of the rose-pink Pogonia's fragrance, and says in his notes in *Summer*, on June 21, 1852: "The adder's-tongue arethusa smells exactly like a snake. How singular that in Nature, too, beauty and offensiveness should be thus combined!"[2] On July 7, 1852, he again mentions these species of orchids: "The very handsome 'pink-purple' flowers of the *Calopogon pulchellus* (now known as *Limodorum tuberosum*) enrich the grass all around the edge of Hubbard's blueberry swamp, and are now in their prime. The *Arethusa bulbosa*, 'crystalline purple,' *Pogonia ophioglossoides*, snake-mouthed (tongued) arethusa, 'pale-purple,' and the *Calopogon pulchellus*, grass-pink, 'pink-purple,' make one family in my mind (next to the purple orchis, or with it), being flowers *par excellence*, all flower, naked flowers, and difficult, at least the *Calopogons*, to

The Thompson Brook, East Pownal, Vermont.

They left their home of Summer ease
Beneath the lowland's sheltering trees,
To seek, by ways unknown to all,
The promise of the waterfall.

Whittier

preserve. But they are flowers, excepting the first, at least, without a name. Pogonia! Calopogon!! They would blush still deeper if they knew what names man had given them."[3]

The Pogonia seems to bloom slightly in advance of Limodorum, and is a delicate, waxen-pink flower. It raises its single terminal blossom about six inches high amid the tall grasses of the swampy meadow. It is not so beautiful as its comrade species, the Grass-Pink; but to me it is sweetly fragrant, and since it is an orchid, it is precious, although small and somewhat unsightly in its suggestiveness.

There are two leaves: one, oblong and sessile, appears in the middle of the stem; and another smaller, bract-like leaf is found at the base of the seed-capsule, bearing the nodding blossom with its alert bearded petals. The roots are little clusters of fibrous threads, loosely attached in the moss-grown mounds of the primeval forest stumps,—which are slowly decaying below the soil in these aged swamps.

The Grass-Pink (*Limodorum tuberosum*) is much more attractive, with its rose and pink-purple blossoms. The spike, often a foot high, bears from two to fifteen beautiful and slightly fragrant flowers. The origin of the generic name, *Limodorum*, comes from the Greek, meaning "a meadow gift." These flowers, according to Mr. Coleman, are called Grass-Pinks in Michigan, while Thoreau also called them by the same name in Massachusetts.[4] The labellum seems hinged at the insertion, and is bearded with yellow and purple hairs. There is seldom more than one freshly blown blossom on the stalk at a time, and thus the plant remains attractive for some days. Beginning at the lowermost bud, each one takes its turn in unfolding, the spike slowly lengthening while the buds constantly increase in size and color.

One interesting peculiarity of this species is that it remains as Nature originally intended all species of orchids,—with the labellum as the *upper* petal, instead of the lower, as seen in all other native species. It will be observed in species of the Orchid Family that a twist of the seed-pod has taken place: if not a complete revolution, at least half a turn. The labellum is, therefore, directed forward on the lower or inferior side, as in the species of *Cypripedium*, where it appears in the position of a shoe or moccasin, instead of holding itself above like a dome, as originally intended by Nature. Darwin says of this: "An enormous amount of extinction must have swept away a multitude of intermediate forms, and has left this single genus, now widely distributed, as a record of a former and more simple state of the great Orchidean Order."[5]

The ovary of the Grass-Pink is straight, and the labellum so hinged that it falls down like an arch above, bearded with delicate hairs. The column bearing the anther, containing four soft pollen-masses, curves slightly at the end, producing a hollow wherein lies the pollinia. The stigmatic surface lies still farther toward the centre of the column. An insect sipping nectar from these flowers, safely enters without

The Grass-Pink. (*Limodorum tuberosum.*)

This is a strange, beautiful orchid with a straight seed-pod (ovary), which causes the labellum to remain on the upper side of the inner whorl, instead of the lower side by torsion as in nearly all other orchids.

distributing the adhesive pollinia, since the anthers containing the cells are so hinged that not until he turns to leave the heart of the flower does he swing open the lid of the cup containing the powdery gold, which fastens to the velvet of his coat beneath his body. The next flower of this species, therefore, becomes fertilized properly, and in turn unlocks her treasure-store as the insect backs off the keel of the pollen mass. Professor Meehan writes that this plant "rarely fails to produce perfect seed-vessels. Yet it is seldom that plants which depend on insects for their supply of pollen, as these are supposed to do, and which are not fertilized by their own pollen, produce seeds from every flower."[6]

It is said that the twisted ovary seen in orchids came about through necessity in fertilization. This has caused, as Darwin says, "the labellum to assume the position of a lower petal, so that insects can easily visit the flower; but from slow changes in the form or position of the petals, or from new sorts of insects visiting the flowers, it might be advantageous to the plant that the labellum should resume its normal position on the *upper* side of the flower."[7] In the present position of the labellum of Cypripedium we observe the convenient resting-place for the bee as it alights and descends to the interior, where are stored the nectar and attractive colors. The insect must be persevering indeed to win the soul of the orchids, since Nature has constructed their organs with such care and modifications. The hidden hinge to the cups of pollen—as instanced in the flowers of the Grass-Pink—demonstrates that even the finest hairs and tissues in these plants have their meaning and their values.

Self-fertilization seems impossible to the Rose-colored Pogonia, which bears but one flower. The plants must inter-cross. An interesting account of the fertilization of this orchid is given at length by Dr. Samuel H. Scudder,[8] in the Proceedings of the Boston Society of Natural History.

Notes:

[1] T. W. Higginson, *The Procession of the Flowers*, p. 21.
[2] Thoreau, *Summer*, p. 198.
[3] *Ibid*, p. 347.
[4] Thoreau, *Summer*, p. 347.
[5] Darwin, *Fertilization of Orchids*, p. 226. 1895.
[6] Thomas Meehan, *The Native Flowers and Ferns of the United States*, p. 104. 2: 1878.
[7] Darwin, *Fertilization of Orchids*, p. 284. 1895.
[8] Dr. S. H. Scudder, *Proc. Soc. Nat. Hist. Boston*, 9: 1863.

8
A Colony of Ram's-Heads in Witch Hollow

The solemn wood had spread
Shadows around my head,—
"Curtains they are," I said,
"Hung dim and still about the house of prayer";
Softly among the limbs,
Turning the leaves of hymns,
I heard the winds, and asked if God were there.
No voice replied, but while I listening stood,
Sweet peace made holy hushes through the wood.
—Alice Cary, *The Sure Witness.*

It was often a temptation during my search for wild strawberries, to saunter through the swampy meadows on the northern slopes of Mount Œta, where nesting bobolinks were busy about their homes. Their happy notes are the first to awaken one in the morning, and almost the last heard at twilight, about the edges of the road and the orchard, where they come in a very business-like way to search for food, crying the while, "Bob-o'-link, bob-o'-link, spink, spank, spink; chee, chee, chee!"

As twilight deepens and the moon comes up from behind the grim form of the Dome, the mournful notes of a distant chorus of whip-poorwills begin, echoing on until far into the early morning. The other noon I was startled to hear a baby whippoorwill practising his melancholy tale on the hillside above the house, where no doubt his mother had lost him the night before. He had "stayed out all night," and knew no better than to sing in the daytime. I suppose his mother had not yet taught him when and how to sing, for he could only lisp now, saying "'T is-so-still! 'T is-so-still!" It sounded very odd at noon, although it was dark and rainy. I searched through the daisied meadow for him, and found that he was a full-sized bird,—too large to be lisping such baby notes, though not old enough to find the way to the twilight woods alone. Perhaps he was backward in his singing lessons, and his mother had punished him by leaving him to practise all day, when other birds of the night were drowsing under the shelter of old logs in the deep wood. So he sang on and on, at intervals, all the afternoon in the rain, out on the grassy hilltop.

I found a bobolink's nest low in the swamp meadow, near where there were many busy "Roberts of Lincoln." Their rich, energetic, gladsome song was very contagious, and brightened many an hour when I was housed, or sat on the porch, watching the storms come up in the north and west.

Mount Œta is one of the foothills of the Dome, lying just west of the Domelet. The Hoosac glides around its "dug-away" base, passing through the narrowest portion of the valley near the Massachusetts State Line. This pass is often called the "Golden Gate," likened to the Pass of Thermopylæ, among the mountains of ancient Greece. Indeed, the warring history of this valley may be comparable with that of the plains of Marathon and the mountains of Hellas. Through the Hoosac Pass, during the French and Indian Wars, have marched the French cadets and cunning Indians, led by General Rigaud de Vaudreuil, to storm and capture Fort Massachusetts near the base of Greylock's Brotherhood. Here they fought, sixty to one. These encounters were but forerunners of the Bennington rebellion among the Green Mountain Boys, and the conflicts at Ticonderoga, which led to one of the world's great battles, fought among the hills and vales of Saratoga.

The summit of Mount Œta is crowned with luxuriant farms, with flowing fields of grain and grasses. Miniature hills and vales between, with little streams leading down the slopes, perfect an ideal pastoral dream. There is none of the boldness in the scene from this height, as observed from Mount Greylock, Mount Anthony, or the Majestic Dome.

Very often the highest summits, especially those of the Dome and the Greylock group, are draped with rosy-tinged clouds and lowering veils of mist at the sunrise hour. One of the rarest visions seen from our modern Mount Œta occurs about six o'clock in the morning, frequently during the months of June and July, when the whole valley of the Hoosac appears filled with a perfect sea of billowy fog, the distant blue mountain peaks rising above. With the golden lights of dawn falling upon this ocean of beauty, one can trace twenty miles of fairy-sea, as the foaming fog follows the serpentine windings of the Hoosac from its source under Greylock, ever broadening toward the plains of Hoosac Falls and the hills of Saratoga. Before ten o'clock the mist usually dissolves, or rises as the sun burns forth.

In all my wanderings, I had kept an eye out for the leaves and seed-capsules of the Ram's-Head Moccasin-Flower (*Cypripedium arietinum*), and had revisited the Amidon Woods, where Lorenna found the first specimen for me, but without discovering any new plants. On Sunday, the second day of July, a friend and myself drove to Pownal Centre. We returned by the Gulf, or "Witch-Hollow" path,—a cross-town road seldom travelled, although shaded and pleasant.

The Perry Elm, Marking the Site of Fort Massachusetts, on the Harrison's Flats, North Adams, Massachusetts, Showing Saddleback Mountain in the Distance.

Here the sounds of the winds, breathing and reverberating through the narrow vales, then dying mournfully in the distance, intimidated the early settlers, who, being superstitious, attributed the sounds to the witches so prevalent in the history of New England. To-day there are no more dreadful sounds in these glens than the hoots of owls and the piping of frogs in the Chalk Pond pools.

We were nearing the pond region. Just west of the road there is a beautiful, ever-bubbling spring, known far and wide to tourists sauntering to Mann Mountain beyond. From this I wished to get a draught of delicious water for my friend, so I hitched the old horse to a tree by the roadside. Somehow this morning I lost my bearings, and entered the wrong ravine. I had supposed that I could find the spring in the dark; but I penetrated the thicket a little north of the right place, by the slab-bridge where, in rainy seasons, the water drains from the hills. Hunting around, however, to learn where the spring lay, I stumbled straight upon a little company of Ram's-Head Lady's Slippers. In my pleasure and excitement, I exclaimed, "Here are Ram's-Heads!" frightening my friend so that she ran clear out of the thicket. She soon returned, however, when assured that there was no danger, and admired the rare little flock with me. There were only a dozen plants in the group, none of course in blossom; but several bore plump seed-capsules, proving that they had bloomed early in the season. I determined to return to this nook another day.

The next morning I started off cross-lots, over the hills afoot, to my sylvan shades, carrying my usual basket and kit of tools, with an added two-quart pail, which I promised to fill with raspberries. These berries were plentiful, I had observed, through the John-Fallow sheep pastures. Here I found a spring trickling from the shelving slate rocks, and this guided me through a meandering network of swamps, all the way to Cold Spring, in Witch Hollow below.

Major frisked about among the fields, and we had a happy time sliding down the dry and slippery pasture slopes. There, at the foot of the hill, we entered a deep, dark woodland, just Major and I, who are faithful, congenial comrades. My constant hound is ever ready to follow my footsteps, and if he chances to lose me, I soon hear his yelp on my track. Dear old Major! I value you more than I tell you by these gentle strokes,—you, whose searching instincts would find me out wherever I might be, and whose keen scent of danger is my constant protection!

Everything was still in the hollow to-day, save for the croaking of the bull-frogs and the buzzing of flies and humming of bees, echoing from the pools and numerous flowers of Solomon's Seal along the edges of the swamp. It was noon when I reached my colony of Ram's-Heads, and I was glad to be sheltered in these cool glades this sweltering July day. I took note near what species of trees my rare Cypripediums grew, and found that they were rooted in loose leaf mould, from long decayed heaps of pine branches and tree-tops, left by the woodman

The Small Round-Leaved Orchis. (*Habenaria Hookeriana.*)

This species is closely related to the Large Round-Leaved Orchis (*Habenaria orbiculata*) and *Habenaria oblongifolia*, with which it grows in company.

when the forest was first hewn from these slopes. Here, also, stood crumbling stumps, and prostrate trunks lay at full length, decaying in the marl and peat. Among this mouldering soil was a pile of four-foot white birchwood—near some of the best plants of Ram's-Head, three of which bore maturing seed-pods. Directly through the group, a wood-path wound around the hill from Cold Spring toward the north, worn by the small wild animals of the forest.

Just east of the plants I had found on Sunday, I discovered at least fifty more, withdrawn to themselves, in aristocratic exclusiveness. I lifted three of the oldest and largest plants, two of which bore large seed-pods, taking them up carefully and with plenty of soil, so as not to disturb the fibrous roots. The layer of leaf mould was loosely strewn, and not so deep here as I had expected to find it. Scarcely three inches beneath the surface, I came to a bright, whitish gravel. The spot was situated on a sloping hillside, which seemed to surround a hollow among the hills, where a glacial lake had formerly slept. It is called to-day "Chalk Pond," the water being whitish at times in the streams flowing from the heart of the region. The soil was rich with unfathomable depths of peat and marl in the lake bed below. Peat is formed by decaying moss, ferns, and vegetable matter in general, while the marl, which lends a chalky appearance to the water and gravel, comes from the crumbling and decayed shells abounding in the soil. This loam seems to be valuable, and the pond bed is now well drained for the purpose of selling the substance as a fertilizer for lawns.

White birch, chestnut, pines, and nearer the pond meadow below, beautiful elms towered skyward. From this corner I searched the hillsides to the north, along the path. At the feet of some chestnut saplings, I found the Small Round-Leaved Orchis (*Habenaria Hookeriana*). The plant was young, and apparently had not put forth blossoms this summer. They appear in early June in this region. Leaving the plant to study another year, I sought the southern hillside, and came suddenly upon a sight which I shall not soon forget. Before me stood the Great Round-Leaved Orchis (*Habenaria orbiculata*), with its two huge, round, flat-lying leaves of a soft emerald green, about eight inches long by seven wide. It bore a tall, bracted spike of greenish-white flowers,—strange, fantastic shapes, trimmed with spurs and hoods and capes. This spike of flowers rose straight up from between the two round basal leaves. It was about two feet high originally, but had been broken, doubtless by the hailstorms of June. The common names of the Round-Leaved Orchises hereabout are "Shin-plaster" and "Heal-all," since they are applied to bruised shins, and are used as plasters for weak lungs. Thoreau, in *Maine Woods*, gives even larger dimensions of the Great Green Orchis found by him in the vicinity of Mud Pond, Moosehead and Chamberlain Carries, Maine,—where he reported it very common in July.

I sat for some time admiring this weird plant; when finding that it had sown seed the former season, I decided to transport it to a garden of civilization, to see if it would take kindly to cultivation. Then I turned westward, following the sluggish yet sparkling stream down from Cold Spring. At times the stream was almost hidden by moss, through which it crept slowly.

This brook enters a large, open, meadow marsh,—the ancient lake bed of which I have spoken before,—the Chalk Pond hollow. Since it is now drained, it appears to be a promising soil in which to seek the Purple-Fringed Habenarias in the proper season. I found the leaves of a plant which I believe to have been that of one of the Purple-Fringed Orchises, but from its producing no flowers this season I was not able to designate it. Here, also, small ferns and luxuriant brakes were sheltered amid the low sumach bushes and willows. Wild grape-vines entwined the trunks of trees, reaching far into the tops of the high elms. One immense elm had been blown over by some northeast hurricane, which had quite recently swept through this hollow. The upturned roots of this rained tree had apparently grown about a deeply buried fellow in the peat and marl, for they still retained the impression of the buried trunk about which they had clung. In the mud and water from which the tree had been torn, lay in its deep grave this log, bare of its outer bark, but still sound and round. It was now well water-soaked, after having been so long sealed from the air and light beneath the earth. How many centuries it had been buried there, no one can guess. The now apparently aged elm upon the surface had torn up several feet of earth as it fell. Forest after forest had thus fallen, a new one rising over it, eventually to give place to another, and itself to form a strata of mould, enriching the soil of these bogs which yield so many floral treasures.

I did not remain in this meadow long, as it proved still too damp to walk through grasses and sedges without water-tight boots. Coming out of this place at the foot of the little ravine below the colony of Ram's-Heads, I ran upon numerous oblong, waxen, green leaves, which at first reminded me of the similar leaves of the Pink Moccasin-Flower (*Cypripedium acaule*). But on closer search for their seed-capsules, I found the fresh bracted processes of a spike containing several ovaries instead of one, as in the Moccasin-Flower. Evidently this plant was not a species of Cypripedium; and although the scape was broken, enough of the alternating process of twisting ovaries remained to assure me that I had found a colony of the early and Showy Orchis (*Orchis spectabilis*), which is one of the first species of orchids to bloom in New England. Indeed, it is said to open the orchid season as early as May 19th, and is found with the Wake-Robins and Arbutus, when the woods are otherwise bare and brown. I secured three of the finest plants.

My basket was now laden with choice species, including those of the Ram's-Head, the Showy Orchis, and two species of Habenaria, a sister genus of *Orchis spectabilis*. The locality had proven a treasure-ground to me, for here were both the Great and the Small Round-Leaved Orchis (*Habenaria orbiculata* and *Habenaria Hookeriana*); while the Tall Green Orchis (*Habenaria hyperborea*) dwells in the deeper bogs along the stream.

The leaves of the Purple-Fringed Orchis (*Habenaria grandiflora*) are hidden in the borders of the open meadow. I found a few plants of that very rare orchid called Adder's-Mouth (*Achroanthes unifolia*), seldom if ever before collected in this town. The plants are so small and inconspicuous that one may search long without seeing them. Two stood among the select company of Ram's-Heads, while others grew along a damp, silent brook bed that had ceased to flow,—a ravine formed during spring freshets and melting snows.

This pigmy of the Orchid Family—with its pale and odorless flower and its unassuming habit of concealing itself in the darkest recesses of our forests and swamps—grows plentifully in its native haunts to the north.

I had searched long and closely for the last month, hoping to find the Large Purple-Fringed Orchis. Thoreau says: "It is remarkable that this, one of the fairest of all our flowers, should also be one of the rarest—for the most part not seen at all. ... The village belle never sees this more delicate belle of the swamp. ... A beauty reared in the shade of a convent, who has never strayed beyond the convent bell. Only the skunk or the owl, or other inhabitant of the swamp, beholds it."[1]

The Yellow-Fringed Orchis follows later, blooming through August and September,—the blossoming season of the flaming Cardinal-Flower, whose brilliant coloring brightens the dark shades along streams in moist woods. The Yellow-Fringed Habenarias are found growing with the Pitcher Plant, and often fill the sphagnous swamps with a glowing mass of orange-flamed torches. Gray considered them among our handsomest species of Habenaria. They are abundant in swamps about New Haven, Connecticut, while the White-Fringed Orchises seek the coast-lines of Massachusetts, although also found sparingly in the highlands.

Species of *Habenaria* are called False Orchises, while species of *Orchis* are known as True Orchises. These species are members of sister genera, but all belong to the Orchid Family. There are but three True Orchises found on the continent north of Mexico, while not less than forty-four species of *Habenaria* are reported for the same area.

The genus of True Orchises comprises eighty species, distributed throughout the temperate zone of the world; while of *Habenaria* there are about five hundred species. *Orchis spectabilis* and *Orchis rotundifolia* are found in Vermont. The latter is the rarer, and limited in its range from northern New England to Greenland. The *Orchis spectabilis* ranges from Ontario southward to Georgia. The third species, *Orchis aristata*, is endemic to the wooded regions of Alaska.

Our common Showy Orchis resembles the Early Spring Orchis (*Orchis mascula*) of England, which Darwin never tired of praising. The high organism distinguishes species of this genus as True Orchises.

The Showy Orchis. (*Orchis spectabilis.*)

Showing the plant nearly natural size. This species is closely allied with the Early Spring Orchis (*Orchis mascula*) of England. It is the most highly organized of our native orchids.

The origin of this distinction lies in the complex structure of the organs of fertilization. The stigmatic lobes, or female organs, and the anther containing the pollinia or male substance in fertilizing, are enclosed in this genus in a pouch or hooded fold above and within the anterior portion of the orifice of the spur. In False Orchises, the stigma and anther are naked, and their glands are exposed. They are also known as Naked Gland Orchises. The more complex the structure, the more highly organized becomes the species. *Orchis spectabilis* displays a marvellous intelligence in its mechanism for inviting fertilization and cross-fertilization. The enclosure of the glands within the hooded pouch protects the pollinia from rains or improper insects.

The moth finds a resting-place on the petaled platform, while he pushes his tongue and head into the depths of the dainty spur attached to the flower anteriorly. In doing this he forces his forehead against the viscid lobes of the stigma, situated in the back, opposite to the entrance of the spur. In pushing, as he must, to reach the nectar in the twisting spur, he ruptures the interior membrane of the rostellum above the orifice containing the pollinia. Each mass of this fertilizing substance in this species contains viscid disks or handles, fastened with elastic hair-like caudicles attached to the pollinia. When the insect ruptures the cellular tissues of the anther, these disks shoot out of their sockets, and fasten firmly to his head. As he flies away, he possesses one or two pollinia, unique in their completeness. In visiting the next spike of the Showy Orchis, he repeats the insertion of his tongue and forehead in the spur of the nectary. The golden horn of pollinum thus rubs against the viscid surface of the stigma, and fertilization and cross-fertilization are brought about. The insect thus accomplishes all that Nature has designed for the future of the species, even if only a small portion of the pollinum is absorbed by the attractive surface of the stigma. One mass fastened to the head of a moth would, in this manner, fertilize several flowers.

According to Darwin, orchids with short-spurred nectaries are fertilized by bees and flies; while those with long spurs are visited by moths and butterflies with long proboscides.

The structure of various species calls for special insects to fertilize and cross-fertilize them. The failure to attract the proper agencies has led Nature slowly to change the organs of many orchids so that self-fertilization might be accomplished. In this way, "an enormous amount of extinction" must have taken place. A wide gap of obliteration intervenes between species of *Orchis* and *Cypripedium*, the former being the most highly organized and the latter the lowest, or abnormal species of the Orchid Family.

The species included under the great genus *Habenaria* grow more abundantly than any other on our continent. It is not unusual to find five or six species of this genus in a neighborhood such as the Bogs of Etchowog or Witch Hollow region. In the latter locality I found four species of *Habenaria*, two of *Cypripedium*, one of *Achroanthes*, and one of *Orchis*, making in all eight rare species for a very small area of swamp-land.

Soon after I reached home with my basket of roots, the front porch exhibited a long row of pots and tin cans, where stood my transplanted treasures, ultimately to be placed in the garden of a friend in New Haven.

It has always been a source of wonder that Thoreau did not find more species of the Orchid Family in the conifer swamps in the Maine woods. His journeys made in July, through pine and cedar and mud-pond regions, should have led to the discovery of more species than he mentions. He writes of but three species of *Habenaria*, one of Ladies' Tresses (*Gyrostachys*), and one of Twayblade (*Leptorchis liliifolia*). To be sure, he found the Great Round-Leaved Habenaria and the two Purple-Fringed species in abundance, but there is no record of a *Cypripedium* in his data save as reported for Concord.

Species of *Orchis* and *Habenaria* are among the oldest orchids known in the records of ancient herbalists and naturalists. Both of our native Purple-Fringed Orchises (*Habenaria grandiflora* and *Habenaria psycodes*) are closely allied with *Orchis morio*, found so abundantly in the fields of England. Pliny, in the time of Christ, knew this plant as *Orchis* or *Serapias*, which Fée has identified with the *Orchis morio* now known in Europe. This species is more nearly related to our Small Purple-Fringed Orchis than to the larger species.

The origin of the name *Orchis* arose from the ancient lore of classical mythology. Orchis, a son of a rural god named Patellanus, failed to observe the rules of politeness while attending a festival of Bacchus, and offended one of the priestesses with his rude behavior. He was reported to the attendants for punishment, who in anger tore him to pieces. His father Patellanus, and his mother, that sweet nymph Acolasia, sought the co-deities' influence, who, it is said, urged the superior gods to command a flower to rise from the earth perpetuating the name and memory of their son. Thus arose the strange untamable species of this family.

The species now known under genus *Orchis* and *Habenaria* had various common names in ancient literature. There were five kinds of Orchis which the Greeks commonly called Cynorchis; this became in Latin *Testiculus canis* and *Testiculus morionis*, and later in England, *Orchis morio*. Satyrion was also an ancient common English name for the species of Cynorchis known to the Greek apothecaries.

In the sixteenth century the Purple-Fringed Orchises of England were known as *Satyrion Royall*, *Noble Satyrion*, *Palma Christi*, and *Royall Standergrasse*. In fact, all species of orchids in 1578 were described under the group of plants designated as *Standlewort*, or *Standergrasse*.[2]

Shakespeare mentions them in *Hamlet* as "Long-Purples" and "Dead-Men's Fingers." Tennyson also speaks of them as Long-Purples in *A Dirge*. Rev. Mr. Ellacomb, in *Robinson's Garden*, alludes to these orchises as "Dead-Men's Thumbs."

From Lithograph in Meehan's *Native Flowers and Ferns of the United States*, I: 1878. By permission.

The Large Purple-Fringed Orchis. (*Habenaria grandiflora.*)

Closely allied to *Habenaria psycodes* of New England and to the English Long Purples (*Orchis Morio*) of ancient literature. They are mentioned by Shakespeare in Hamlet.

There with fantastic garlands did she come
Of crow-flowers, nettles, daisies, and long purples.

Hamlet, Act IV., Sc. 7.

The Great Royall Satyrion of England and Germany, known to Dodoens and Lyte in 1578, was found in meadows and moist woods. The flowers were light purple, and gave forth sweet perfume. The roots were described as double, like a pair of hands, and each palm was parted into four or five small roots like fingers; one palm being withered and spongy, the other full and sound. From this peculiarity of form many of the names were undoubtedly derived. There was also another small purple species of Royall Satyrion, with a perfume like musk. The roots were like the larger purple Royall Satyrion.

The roots of Royall Satyrion were used as remedies against many diseases. "If an inch or as much as one's *thombe* of this roote be pound and ministered in wine, it is good for many diseases," writes Dr. Nicholas Nicols, according to Dodoens and Lyte in 1578.[3]

These orchises have figured in literature from time unknown, and although shy in New England, seeking the haunts of moose and bear, they delight still to grow in hearing of the cathedral bells in old England, where they are the common flowers of meadow and borders of corn-fields.

The proverb, that all things come round to him who waits, may for the orchis-hunter be paraphrased rather, "All things come round to him who tramps." I was destined sooner or later, by lonely lake or mountain bog, to find the Purple-Fringed Orchis for which I had so long searched. Later in the season, on July 8th, I visited Notch Brook, North Adams, a stream flowing down through the northern Notch Valley. Wandering past the beautiful Cascade, I slowly explored the wooded vales among the Ragged Mountains. The afternoon was sultry; the sun pouring down upon the parched sod of the rocky pasture-land had shrivelled up the grasses, and now the bushes themselves were turning brown, and the leaves curling up on their edges. Through the trembling haze, partially due to the vile smoke of civilization, which arose from the various factories in the City, the sun appeared as a round, red ball of fire.

I had chosen a poor day for walking, but there were cool, shady retreats on the way, where I could find rest and shelter. I clambered down from the slopes of Aurora's Hill, into the shadow of the valley's smoke, crossing the sluggish stream of the Ashuilticook, by way of the iron bridge in Flag's Meadows. I climbed to the swamps along the Ragged Hills leading to The Notch. Here the slopes of pasture-lands above State Street are clothed with bushes and brambles, through which rough, stony paths wind, where dwell the children of sunny Italy. Witt's Ledge of lime and marble stone lies along this swell. These rough paths, with wooden steps leading summitward, were new to me.

Upon the brow of the hill was a small pond hidden at the head of an extensive swamp, amid willows and lush tangled grasses, where little lads were bathing. It was one of those wild mountainous pasture-lands where blackberry briars and sweet-fern run riot, and where the pepper-bushes and tall brakes shed forth an aromatic perfume under the full blaze of the summer sun. About the drier portions of the swamp were well-worn cow-paths, winding irregularly about the hummocks and boulders; and along the borders grew many familiar weeds and vines amid the swails and flags.

My boots being high and waterproof, I waded warily through the coarse lush grasses and cat-tail flags, encountering many deep pools. As I pushed forward, my heart sang in the very joy of living:

> Is this a time to be cloudy and sad,
> When our mother Nature laughs around;
> When even the deep blue heavens look glad,
> And gladness breathes from the blossoming ground?[4]

At the west end of the cow-path I came suddenly upon one tall, Purple-Fringed Orchis. This was my first good fortune in finding this beautiful species, although I have since found many. I stood long in wonderment and silent adoration before this fragrant beauty of the weird and lonely bogland, rearing its strange fringed petals high above the common swamp grasses. Searching about to the north and south, I found a colony of six more spikes, which assured me that I would be justified in taking the first plant I had found; and placing it with the utmost care in my crowded vasculum, I then proceeded mountainward.

On the very brow of the hill I wound around to the left, entering the wood-road leading to the Notch Valley. A beautiful cold spring gushes out in the heart of this wood, under the hill at the right, near the Cascade path. I freshened my flowers here, and hurried on to the famous foot-bridge over Notch Brook, plunging on down through the hemlock wood to get a hurried view of the Cascade below.

As I returned homeward over the heated fields, I found the atmosphere very exhausting; and the flowers, although protected in my botanizing can, were wilted. Measuring the broad expanse that intervened between me and the hill of Aurora's Lake, the journey seemed interminable. The distance was finally covered, however, and both my fatigue and the fact that I was late for tea were forgotten in the ecstasy of having found that first Purple-Fringed Orchis. This spike grew, and every bud expanded, until within a few days it became beautiful indeed, giving forth its delicate fragrance, and proving itself the prize I had esteemed it, as I lifted it from the dark earth of the bogland of northern Berkshire.

Notes:

[1] Thoreau, *Summer*, pp. 84-85. 1884.

[2] Dodoens, *History of Plants*, p. 156. 1578.

[3] Lyte's translation of Dodoens' *History of Plants*, pp. 161-162 (r ed., 1578).

[4] Bryant, *The Gladness of Nature.*

The Blackberry Blossoms from Mount Œta, Pownal, Vermont.

9
Over the Huckleberry Plains

Thou shalt gaze, at once,
Here on white villages, and tilth, and herds,
And swarming roads, and there on solitudes
That only bear the torrent and the wind.
—Bryant, *Monument Mountain.*

On July 17th, two days before departing from the Hoosac Valley, I was guided to a group of swamps lying along the summit of the Domelet. The brow of this mountain is yearly devastated by forest fires, after which it appears quite barren, save for the trees and bushes protected in the swamps. A few tall trees, branchless and blackened, stand as sentinels about the huckleberry plains. But soon the young and tender growth of oak, chestnut, and birch springs up on these rocky ridges, while the clearings everywhere are carpeted with low blueberry bushes.

Between the Domelet and the Dome lies a valley known as "Rocky Hollow." The ledges of rock, walling it about, bear deep erosions in evidence of the Ice Age, when a gigantic glacier once crowned and rounded the Dome. The formation of these deep vales lying at its base is due to the moraines which flowed down from the ice-capped heights.

The little swamp-like pockets along the summit of the Domelet, where luxuriant trees locate the moisture of springs, were formed, perchance, when a deeper lake rolled over this peak.

In ascending the Domelet, we drove around the northern brow of the mountain, up by the County Road,—frequently called the "Dummy Road" in Pownal, because a deaf and dumb man once lived in the vicinity. Soon we turned off eastward, beyond the Dummy Farm, through the low bushes, until we came to a shady vale. We unhitched our horses from the wagon, and fastened them to trees; then we proceeded to explore the hills and plains, carrying pails for berries, and a basket and spade for collecting roots. The flora of this region appeared luxuriant all along the road, as well as over the ledges and plains. I found great numbers of plants of the Pink Moccasin-

The Yellow Clintonia (*Clintonia borealis*).
Rattlesnake Brook Region, Mount Œta, Pownal, Vermont.

Flower (*Cypripedium acaule*). The unusually large leaves were of a deep dark green, with marked veining. Many stems bore seed-pods, which were the largest capsules for this species I have ever seen, being an inch and a half long, with circumference in proportion.

The beautiful emerald-green leaves and bright berries of the lily, *Clintonia borealis*, were almost as common as the piles of sphagnum and the tall brakes and ferns on the edges of these swamps; yet everything about recorded the ravages of the recent hailstorms. Very few seed-capsules could have remained to mature their seed this season, as most of the plants were either badly bruised, or broken from the root, causing the ovary of the flower to droop and wither. The low huckleberry bushes, known as the dwarf black species (*Gaylussacia dumosa*), were also damaged by the hailstorms, and were without fruit.

We came upon numbers of trees shattered by lightning, and blackened pine stubs and "turnovers" mingled among the beautiful evergreens of the tangled swamp. Low blueberry bushes, with rich heaps of ashes about their roots, covered the rolling, rockbound plains, as far as one could see. Huckleberries usually thrive in the trail of forest fires. Indeed, the spring and autumn fires are often started by the huckleberry venders for the sole purpose of securing a better yield of fruit for supplying the market. These berries are among the small fruits which have not thus far taken kindly to cultivation, as has nearly every other wild berry in the markets to-day.

We found the third swamp eastward marked by the odd spires of the Scrub Pine (*Pinus divaricata*), and the Red Pine (*Pinus resinosa*), which is often wrongly called Norway Pine. These evergreen trees were known to Theophrastus before Christ. There were two kinds, the wild and the garden trees. Many species of each are described, the pines and spruces not being distinguished from each other.

I observed also many Dwarf Black or Double Spruces (*Picea Mariana*),—very dark green trees with pretty cones. The name for this tree originated with Theophrastus. It became in the Latin *Pinus Mariana*.

Frequently I saw a lone Balsam-Fir tree,—*Abies balsamea*. The name *Abies* comes down to us from remote antiquity, since this tree grew in Greece, and was valued by the learned physicians before Christ for the balsamic resin found in the bark of young trees. Matthiolus and Peter Bellon described this substance as bitter and aromatic, similar to citron-pills. In England, this resin was known to the writers of the sixteenth century as the Turpentine of Venice. In Canada, where this tree is abundant, it is called "Balm-of-Gilead Fir," or "Canada Balsam." It is common on the summit of the Dome. The powerful balsamic fluid drawn from it is now used medicinally. This species resembles the black spruce, save that it is of a silvery-green color, giving forth its peculiar fragrance, and producing small blisters on its trunk and branches, which the spruce does not.

Cedar is not plentiful in the Hoosac Valley region, our only species being American Arbor Vitæ (*Thuja occidentalis*), often called white cedar. Northward this tree forms extensive cedar swamps, which are rich haunts for species of *Orchidaceæ*.

On the border of the third swamp, and in the heart of it as well, grow High Huckleberry bushes (*Vaccinium corymbosum*). Blueberries are known in New England as huckleberries, and this common swamp species grows very tall. These bushes before me were over twelve feet in height. The Dwarf Low Blueberry (*Vaccinium vacillans*) grows from one to two feet high. The Early Dwarf species (*Vaccinium Pennsylvanicum*) is from six to fifteen inches in height, and produces our earliest market blueberries or huckleberries. The late Dwarf Low Blueberry ripens late in July.

The giant bushes in the swamp were laden with both green and ripe fruit. The cadet-blue berries hanging side by side with the soft velvety crimson-purple fruit of Shad-trees (*Amelanchier Canadensis*) made a pretty dash of color among the rich greens. Most country lads are familiar with the mountains at this season, and to go "shad-berrying" is one of their pleasures. In Pownal one hears of shadberry pies and cakes with happy anticipations.

These berries are fresh and sweet, eaten direct from the bending branches, but they become as bitter as medicine after being gathered for any length of time. Their white flowers often appear early in April and May, and brighten the waste places along with the Pigeon Cherry blossoms,—better known as Wild Red Cherries (*Prunus Pennsylvanica*). The flowers of the latter tree are also white, producing small light red cherries, which delighted flocks of returning pigeons before their extermination. The generic name *Prunus* is the Latin for plum or prune, derived from the Greek for all species commonly known as Sloes, Bullies, and Snags. The species of *Prunus* and *Amelanchier* are members of the Rose Family, as their miniature rose-blossoms indicate.

The original designations of huckleberry, or whortleberry, are also of ancient derivation. The species of *Vaccinium* were known to Virgil under the title of *Vacinia*,[1] because their berries were little. The ancient writers recognized the black, white, and red fruited species. The white was seldom seen, however, while the red also was rare. The true English name for these berries in the sixteenth century was "whorts" or "whortleberries." The black whorts grew commonly in many woods in England, in June and July.

After wandering through these swamps on the Domelet for two or three hours, and securing some fine roots of the Pink Moccasin-Flower for the New Haven garden, we slowly walked back toward our horses in the shaded vale, up and over ledges and rolling hills, passing a ridge of outcropping marble. We finally sat a while and drank in the cool mountain breeze, catching here and there through the trees the varied panorama of the great world below and the clouds above us. Distant sounds from human abodes rose to our ears faintly,—such as the engine whistle of the "Wild-cat" Express as it wound through the deep-cut valley of the Hoosac, nearly a thousand feet below us, westward beyond Mount Œta. On one of these marble ridges, along the plains, I found several plants of the Large Round-Leaved Orchis (*Habenaria orbiculata*).

We drove homeward by way of White Oaks Road,—southward along the entire summit of the Domelet,—getting an excellent view of the Hoosac and Green Rivers, following their serpentine windings about the hills and vales far below us near Williamstown. White Oaks is a remote corner of Pownal, lying, however, partly in Williamstown in northern Berkshire, a region locally noted for the earliest arbutus blossoms.

In nearly all the swamps I have visited, I have found a long procession of flowers marching close upon each other through the seasons,—from the trailing arbutus and the snowy dogwood blossoms in early April, to the golden-rod and asters of the late October days. Even as late as December 8th I have found the dainty dandelions and violets running wild with glee, only to be frozen before sunrise the following morning. It can be said that in some seasons different flowers bloom nearly every month, in the Hoosac Highlands, if the *Transcript's* reports be true.

"On February 1, 1900, some trailing arbutus was brought from the woods. There is usually a little strife in the spring for the distinction of bringing these first flowers, but Mr. Briggs has forestalled all the flower hunters this year by his January discovery, which is most unusual."[2]

I have collected March arbutus in the White Oaks as early as the 12th, although never in January or February. Indeed, there come many arbutus days long before April and May, if only we go abroad to realize them in the warm, sunny glens among the Taconic Hills, where the cold winds never blow in March.

Notes:

[1] *Vacinium* comes from *Baccinium*, and was derived from *Baccæ*—Dodoens, *Hist. Pl.*, 1578.

[2] Clipping from *The Transcript*, North Adams, Mass., Feb., 1900.

Second Season

White, innocent twigs of apple, idly swaying,
Shed a suave fragrance on the flattering breeze.
John S. Van Cleve.

10
Westville Swamps and Mount Carmel, Connecticut

> When, formerly, I have analyzed my partiality for some farm which I had contemplated purchasing, I have frequently found that I was attracted solely by a few square rods of impermeable and unfathomable bog. ... That was the jewel which dazzled me.—Thoreau, *Excursions*.

May the 1st I departed from New York, to find in bloom many of the earlier flowers that I had missed last year in the Hoosac Highlands. I followed much the same route through Connecticut as I had taken the season previous. The country was aglow with the subtle breath of spring sunshine, that inspires the soul of earth to rise and sprinkle her fields with pulsating life and song. I started out alone to explore the Bogs of Westville, where the dainty Grass-Pinks and Pogonias would later bloom.

There is much diversity of soil about New Haven: it proves a meeting-ground for Southern and Northern species of plant life. The swampy regions of the Great Salt Meadows produce a foreign vegetation that emigrates to our shores; while the rocky ridges of the hills about the City furnish hiding ground for rare ferns and flowers found far northward.

I rode to the end of the car line, near which I turned off into a thicket, and over a bridge above the milldam. On either hand broad fields of marsh-land stretched out to meet the low, rolling hills. To the right, a path led up the slopes of a cow-pasture, along a little stream to West Rock. The damp hillside was carpeted with Innocence or Bluets (*Houstonia cærulea*), and numerous colonies of violets; and amid the moss-grown hillocks, in the woods, the Dog's-Tooth Lily (*Erythronium Americanum*) nodded its yellow bell. This lily, so long designated Dog's-Tooth Violet, is a plant having a broad continental range,—found from Nova Scotia to Florida, and from Maine to Arkansas. In the South it blooms in March, in the North in May. It grows not only in the low coast hills and valleys, but is known to thrive at an altitude of 5500 feet in Virginia and in the North.

The leaves of Dog's-Tooth Lily resemble a species of Orchis known to Dioscorides as *Satyrion Erythronium*, from which Linnaeus in 1753 coined the present generic name *Erythronium*, signifying "red," for the genus of Dog's-Tooth Lilies. Wherever the appended term of "Violet" originated for these lilies is not known. According to Frederick H. Blodgett: "In one of the old botanies in the library of the Agricultural Department (Washington), there is a colored plate, illustrating the European species with the name *Viola dens-canis*, with pen notes, giving the later and more modern names also."[1] That the plant was always considered a lily, however, instead of a violet is evident.

Dr. Rembert Dodoens, as early as 1578, thus described it: "This low base herbe, hath, for the most part but two leaves, speckled with great red spots, betwixt which springeth up a little tender stalke or stem with one floure at the top hanging downeward, which hath certain small leaves growing together like an arch or haute, and like the wild lily." (The *Amarillis* of the Spaniards.) "The names of this herbe now are called *Denticulus canis* and *Dens caninus*, and others call it *Satyrion Erythronium*, wherewithall notwithstanding it has no similitude."[2] It was known to Dioscorides as *Lilium sylvestre*, and Dodoens remarks that "It may well be called such," since the flower, when "it hangs downeward toward the grounde, is much like the wild lilies, saving it is smaller."[3]

Dioscorides (23-77 A.D.) knew this plant as *Ephemeron non lethale*, which was also known in Latin as *Lilium sylvestre*.

Dodoens therefore wrote that if Dog's-Tooth be *Ephemeron*, as it seemed to be, the essence extracted from its root by boiling water, according to Dioscorides, was good for the teeth. But the name was more likely suggested by the fact that the bulbous root is shaped like a canine tooth. The appended "violet" originated, perhaps, with children, since this lily blooms in early springtime with violets, bluets, marsh marigolds, and arbutus.

The names of all plants among the early Greeks and Romans originated from the shapes of the flowers, leaves, or roots, and also from their medicinal properties. Dioscorides knew another species which he designated *Satyrion Erythronium* or *Dioscorides Satyrion*, signifying Red Satyrion,—known to the ancient Syrians. Satyrion was the ancient "shop" name for species of Orchis.

Without doubt our generic name for Dog's-Tooth (*Erythronium*) is a corruption from the Red-spotted Satyrion, whose leaves became confused with *Lilium sylvestre*. Burroughs has remarked that bulbs of lilies in general lie near the surface of the ground. The bulbs of *Erythronium* are often found at a depth of eight inches or more in the earth, however, according to the age of the lily. The young plant often produces but one leaf, and its bulb is loosely attached to the moss on the surface, while the older plants produce two leaves, their bulbs each season sinking deeper into the soil.

Many common names have been suggested by botanists to replace the seemingly inappropriate name of Dog's-Tooth Violet. While the appended Violet is misapplied, as we observe, the name Dog's-Tooth is of ancient origin, and really has more appropriateness than does the generic name *Erythronium*, since none of our species produce red flowers. The name Dog's-Tooth, therefore, was purposely dropped in the *Illustrated Flora* of Northeastern North America, and the name Yellow Adder's-Tongue substituted for the species. There is also an Adder's-Tongue Fern (*Ophioglossum vulgatum*). We therefore see that the re-establishing of the former and ancient name, Dog's-Tooth Lily, for species of *Erythronium* is to be preferred, not only according to the moral rule of priority, but because it is actually the legal common name.

The Westville Swamps were sparkling with these yellow lily-bells, while in the woods along the sluggish stream, the Marsh Marigolds—often called American Cowslips—were holding up their golden goblets to be filled with morning dew. Farther up the stream, near a rude plank bridge in the pasture roadway, I found a baby turtle basking in the sunshine. He was no larger than the hollow of my palm. The little fellow was too frightened to tumble off his stony couch and run for the stream. He sat still and eyed me distrustfully. He drew in his head and toes, and I lifted him gently in my hand, placing him in a paper bag among the flowers I had gathered. I intended him for a surprise in the school aquarium.

Climbing far up the side of West Rock, I looked over the Woodbridge fields and toward West Peak, near Meriden. In the dim distance the Giant's form was outlined against the horizon at Mount Carmel. This mountain assumes the form of a gigantic Egyptian mummy. The hands are folded across the breast, and the head and feet are stretched in stiff dignity—so to remain through the ages.

Among the wooded hills and vales below, the cloud-shadows chased each other to the distant mountains far beyond. That pile of granite upon the brow of West Rock, designated in history as Judge's Cave, stood guard over the hills about me. Farther off, toward Westville, many a roof glistened and peered out among the newly leaved trees of the hillsides. One of them was that of Ik Marvel's home, where perchance he had smoked the dream pipes of his *Bachelor's Revery*. Smoke now curled above the chimney-tops, full of the drowsiness of May mornings.

The song of thrushes and orioles amid the bushes burst joyously upon me, and during the interludes I heard the hum of bees and the distant murmurings of streams. Butterflies sailed by, flashing their brilliant colors in the sunlight, and the air was laden with the delicate fragrance of early woodlands. It was a day marked by hope and promise. Who can forget those fields of spring where forget-me-nots and violets bloom?

During the afternoon of this glorious day, I journeyed to Mount Carmel beyond Lake Whitney. The old canal from Northampton to

New Haven formerly passed along this valley, and although the channel is partly filled in, the towpath still remains, and is well-trodden. I followed it from the end of the car-line, until I reached an elevated ledge of rock to the right. This little hill, clothed with white cedars and junipers, lies beneath the stern brow of the Giant, whose face is plainly outlined against the sky far above the village.

After exploring the ridge hereabout, and finding it covered with Columbine in bud, I descended to the hollow, along the stream. In the rocky crevices at the farthest end of the hill, where the midday sun poured down upon the colony, I found the Columbine flowers in full bloom. I pushed my hand beneath the matted soil, and the plant with roots entire loosened and was easily lifted. Poison Ivy grew about over the rocks, warning me to be cautious. I discovered the Dutchman's Breeches (*Bicuculla Cucullaria*), but they were faded and nearly fallen. I observed here near the Columbine also another species of the genus not often seen in this locality. It is known commonly as the Bleeding-Heart (*Bicuculla eximia*), formerly designated generically as *Fumaria*. It grows in rocky places, especially in southern New York. I collected it once about Montclair, New Jersey, along the Orange Mountains, and on the borders of Bronx Park, New York City, and in the vicinity of Mount Vernon. This species belongs more especially southward, extending to Georgia and Tennessee, flowering from May until September. There are about fourteen species of this genus found in North America and Western Asia. Three of this number are reported for the Atlantic Region of North America. The third species is known as Squirrel-Corn (*Bicuculla Canadensis*), and is very similar to the Dutchman's Breeches, save that the plant is smaller.

While waiting for the car at Lake Whitney Junction, on my return to New Haven, I opened the iron gate and wandered along the wooded edges of the shore. I soon distinguished that homely weed of our old door-yard walls and tin-can waste-heaps—the Cat-Mint or Catnip. I gathered some tender shoots for pussy Yale. The name *Catnip* is of ancient origin, derived from the Latin, *Nepete*, the name of an Etrurian city. Our native plant is of European origin. Dodoens, in 1578, writes of this plant: "In shops it is called *Nepita*; in England *Pep*, and *Cat-Mint*, in French *Herbe de Chat*." The name of our single species is to-day *Nepeta Cataria*. There are about one hundred and fifty native species of this group of plants found in Europe and Asia.

This family of Mints was known to the ancients as *Calamint* or *Calamintha*, and included in Dodoens' day four or five species described by the Greeks, each of them having several names marked by different medicinal virtues. The first kind was called Mountain Calamint; the second was known as Wild Pennyroyal, and the third variety as Cat's-Mint or Cat's-Herbe,—just described as *Nepeta*. The Wild Pennyroyal of the ancients resembled the cultivated species at that time also, which was known in England, during 1500, as Podding-Grasse or Pudding-Grass. Pliny attributed twenty-five medicinal and mystical properties to Pennyroyal, in Christ's day, while Dodoens and Lyte mention fourteen uses of the herb in 1578. Xenocrates prescribed "a branch of Penny-Royall wrapped in wool," and placed beneath the bedclothes as a remedy against malignant fevers.

The Mints in general were called in ancient apothecary shops *Mentha*. This group of plants was known to Theophrastus, and named *Mentha*, from a nymph fabled in classical literature as having been changed into this plant by the jealous Proserpine.

Yale was delighted with the *Herbe de Chat*, and scented spring and the cat's heal-all indeed in these tender sprays, rolling and purring over the leaves like a tiger, until at last, soothed, he fell a-napping.

The next morning, May 2d, I started for the Berkshire Highlands, where I arrived in the afternoon. I remained on Aurora's Hill in North Adams about a week. I was too early for orchid flowers, and the arbutus was still in bloom. These were the real arbutus days here, in spite of their breaking the record now and then by blooming in February. I soon left the slopes of North Adams for Pownal Hills. I found them bleak and cold, and that here, too, I was ahead of all species of the Orchid Family.

I made several excursions to the heart of Rattlesnake Swamp and Rattlesnake Ledge, between May 7th and 15th, searching for Trailing Arbutus. I courted the Swamp of Oracles and the Glen of Comus, watching the buds unfold. The woods were bare and leafless, the paths and dry-brook beds were flooded with sunshine. There is a desolate expression to these deep swamps in early spring, before the tender green leaves of trees and budding flowers burst forth. The birds, however, are here, and their song tells us that it is spring instead of the Indian summer of late autumn. These awakening days of the sleeping woodlands reminded one of the death of the flowers in November. Equally sad is the birth and death of the flowers. The sod steams with the warm southern sunlight pouring upon it until, behold, a week later the green, luxuriant foliage hides all the rocky paths in dense shades, and sprinkles dainty stars and clinging vines over all the ruins of the autumn's faded stalks and leaves. May holds greater charm and more silent mysteries than the overflowing joy of full-grown June.

I wandered on through the winding paths, finding them draped with mosses and Goldthread blossoms and Painted Trillium. I continued daily to search the woods through and through for the first sight of the Pink Moccasin-Flower. I found three species of *Cypripedium* on May 15th, and was a little curious to observe how the race would end in their unfolding. There were the rare Ram's-Head, the Pink and the Large Yellow Moccasin-Flowers in bud. The Pink Cypripedium was the first to open, upon the 19th, while on the same day also the Large Yellow Moccasin-Flower burst into bloom, dropping its long, twisting side petals gracefully beside the stump of a hemlock tree fallen across Ball Brook. The Ram's-Head, which is supposed to be the earliest *Cypripedium* to bloom, was not so fortunate. Venturing where it stood

The Woodman's Road through Rattlesnake Swamp,
Mount Œta, Pownal, Vermont.

These dim-aisled forests His cathedrals, where
The pale nun Silence tiptoes, velvet-shod,
And Prayer kneels with tireless, parted lids.

Ella Higginson

among the Amidon Pines one sunny morning, I found the bud still sheathed in the tender green bract-like leaf, laid low and withering upon the ground beside the ruined stem. I picked it up and wondered what or who could have brought about this tragedy. The bract, containing the bud, appeared cut from the stem as with a keen-edged knife. As I held it, I observed a large-worm concealed within the plicate folds of the bracted leaf, amid the musky sweetness of the bud. It was round and emerald green, and unlike most worms was possessed of a great degree of spryness; for before my senses were astir, it had dropped to the ground, wriggling out of sight among the leaves and earth at my feet. I poked about, hoping to follow its trail, but to no avail.

That this rare *Cypripedium* has a vile destroyer in this worm, is evident. I felt that I should like to know more about the worm's life, and why it seeks the rarest species of *Cypripedium* in North America to feast upon. Perhaps this may account for this flower's rarity in Pownal, as well as throughout its continental range. I have observed the worm among the Ram's-Heads in Amidon's Pines during the past two seasons.

The budded plants of this species in Witch Hollow were also blasted in embryo this season. The two buds observed turned brown and withered on the stem, while yet very tiny, leaving the bract perfectly intact. Some plants, however, were untouched by the destroying worm. They were without doubt too young to bloom, as it requires four or five seasons before seedling Cypripediums produce perfect blossoms in their native haunts.

The seeds of orchids are minute, mostly "spindle shaped," are found in great numbers, and resemble fine sawdust. The ovules of the Orchid Family mature slowly, and as far as scientific observations count, in the tests made with the cultivated species, it is said the seeds are a year in coming to maturity. These fertile seeds need several months in which to germinate, requiring, as will be seen, some years to produce seedlings old enough to blossom from the self-sowing capsules of our wild native orchids. Furthermore, a certain temperature and continuous moisture are absolutely necessary to produce seedlings. Our variable Northern climate is one of the natural causes of hindrance to the production of native seedlings of this sensitive family. Continuous moisture does not prevail long enough to promote perfect germination of the thousands of seeds annually produced. They die in embryo from lack of moisture during dry seasons; or if developed, are frozen out in the ice-capped swamps by our long, harsh winters; or in the tender seasons of their seedling-hood the following years, are dwarfed and die from drought. This, in part, is the explanation of the lack of natural seedlings in our native orchid haunts. Adding to this the destructiveness of man, the beasts of the fields, worms, and storms, and the tardiness of insects in fertilization, a hard struggle lies in the perpetuation of the family's future generations.

Orchids are spoken of as the "weeds of the tropics"; notwithstanding their devastation by the present orchid craze,—which far outstrips the tulip mania in Holland two hundred and fifty years ago,—in regions where continuous heat and moisture prevail, germination of their fertile seeds is rapid and natural.

Notes:

[1] P. H. Blodgett, *The Plant World*, p. 52, March, 1902.
[2] Dodoens, *History of Plants*, 1578.
[3] *Ibid.*

11
May Showers and White Moccasin- Flowers

> If a man walk in the woods for love of them half of each day, he is in danger of being regarded a loafer; but if he spend his whole day as a speculator, shearing off those woods and making earth bare before her time, he is esteemed an industrious and enterprising citizen.—Thoreau, *Letters*.

There is something charming about an unwearied rain in spring. I chose a day upon which rain was falling to journey through the swamps, observing my orchid buds. The clouds would lift now and then with sudden brightening, although the gentle patter of the rain was constant. The wind scarcely stirred the leaves. Nature was quiet in her weeping, as a heart that has a grievance which it does not care to share with any one. The meadows grew green, the buds expanded, and the heart of May began to pulsate and sing new songs.

I started out to visit the Glen of Comus, but found the underbrush too laden with rain. I then decided to go through the fields and seek the Chalk Pond colonies. Over the hills lowered a heavy fog, which, as the rain slackened, would for a time lift again, showing the blue peaks in the distance. I turned westward from the glen, through Amidon's Pines. I had soon passed beyond the limits of this sheltering wood, making a gradual ascent through the raspberry pastures of John-Fallow.

The higher I climbed the harder it poured. However, I arrived among the low white birch saplings and berry bushes. Here I managed to shake the rain off them, becoming as bedraggled as though I had waded in a stream. My umbrella began to leak, and my cap and hair were being soaked, the water actually running down my face. Entering the deeper underwood of birches, I aroused a flock of sheep and their lambs. They ran bleating after me, asking for salt. A mother followed me closely, stamped the earth with her tiny feet, showing her petulance and fear; although she did not turn and run from me as I ventured nearer, but rubbed her nose against my hand.

I now began descending the western slopes of John-Fallow, and was in sight of the woods closing about Witch Hollow. Upon entering the thicket, I soon found my colonies of *Orchis spectabilis*, which were not yet unfolded, although it was May 20th. The Cypripediums had come in far ahead of them this season.

The group of Ram's-Head also disappointed me, the buds having been blasted in embryo. The plants, however, looked healthy and promising.

Chilled through as I was on my way out of the woods, I thought of stopping at the nearest house on Butternut Lane for a drink of hot milk. I refrained, however, because of my fog-covered garments and the curiosity I might arouse in the neighborhood. Onward I trudged another mile or two, up through the pastures, across the old Welch Farm, following the grass-grown road that originally led from Mount Œta to the valley of the Hoosac, during Revolutionary days.

I had been out about four hours, and it was time I sought shelter, since I had waded through the tall grasses and bushes, regardless of the rain upon their leaves. Once in the house, I realized the comfort of possessing warm, dry garments.

On May 23d I made a journey to Rattlesnake Swamp. Arbutus was still in blossom near the hemlocks,—late clusters, indeed, hiding in the moss at the feet of small spruces, where the ice and snow had lingered latest.

The children in District Fourteen delight in surprising me with strange flowers. Among these I frequently find rare species of plants to name and identify for them. A delicate spray of the Purple-Flowered Clematis was brought to me recently. This vine is rare hereabout, growing only in rocky woods about the Rabbit Plain, and along the Gulf Road of Witch Hollow.

Children in the country districts are the first nature students in spring. In May and June the woods and fields become veritable classrooms in which Nature alone presides as instructor. A dense oak and pine forest formerly sheltered the vale near the schoolhouse, where the children seldom dared to wander without their teacher. The wood was dark and full of the twilight shades of the virgin forest trees of our New England hills.

In May, many seasons ago, the Purple-Flowered Clematis (*Atragene Americana*) grew abundant in the heart of this rocky wood, covering, in one instance, a bush six feet high with its graceful vine. This plant is rare from Maine to Minnesota. It ranges northward to Hudson Bay, and southward to Virginia, often ascending great heights. It is reported in the Catskills at an altitude of three thousand feet. There are but three species of this genus found natively in the North Temperate Zone, one being reported for the Rocky Mountain region, the other farther to the northwest coast of America. The common Virgin's-Bower (*Clematis Virginiana*) grows also in rocky places, covering roadside walls and bordering swamps and river banks in July and August.

Later in the autumn, this species is very attractive. The seed-pods burst and produce a light, feathery down—little wings to aid the seeds in their flight, like those of the dandelion and milkweed. The seed-capsules of the Purple-Flowered Clematis also produce tails like the plumes of a feather.

Several species of *Clematis* were known to the ancients in Christ's day. The name originated with Dioscorides, and was used to designate all climbing vines. He knew three kinds under the generic name of *Aristolochia*, named in honor of Aristotle. The "branched vine" with "deepe violet floures," was called *Aristolochia clematites*. Peter Bellon of ye olden time remarks that this plant grew in the mountains of Ida in Crete, or Candie. Carolus Clusius reported it as growing among the bushes and briars about the city of Hispalis, or Civill, in Spain, before the sixteenth century.

I visited Oak Hill Cemetery on May 29th,—a very good place to observe the early flowers of the woods about the valley. The country folk come here with their laurels for Decoration Day, as Milton came to his "Lycidas," "to empurple all the ground with vernal flowers."[1] Here may be seen the pink azalea, the marsh marigolds—those golden-cups of *Caltha*,—violets, and painted trilliums amid the bunches of pink and golden moccasin-flowers, brought here in abundance by the school children.

On June 5th I sought the swamps of Etchowog. I followed down through the Glen of Comus, in search of the great colony of Pink Moccasin-Flowers. I found them in full bud,—two hundred in number, as formerly. As I entered the hollow, I found in the middle of the path a Small Round-Leaved Orchis (*Habenaria Hookeriana*).

This region is being slowly despoiled of its stately pines. I saw fresh scars of the axe among them. Three first-growth trees were laid low, piled on the side of the road.

I followed, as usual, the path through the Swamp of Oracles beside Ball Brook, leading out through the clearings of Ball Farm. Here I waded through Iris Swamp beyond, coming out to the pasture-land of Kimball Farm.

This season, many changes have occurred in the Kimball Bogs, the hillsides closing in about them having been almost sheared of their trees. This results in flooding the heart of the swamp with sunshine, and may in time dry up the growth of the beautiful moss known as Sphagnum, and also destroy the Buck-beans. The cows were browsing among the small tamaracks, and no signs of the Showy Queen of the Moccasin. Flowers were visible hereabout this June. The Tall Green Orchis (*Habenaria hyperborea*) grew luxuriantly in a pool over the fence near the clearing. Purple Trilliums were also very abundant along my path. I passed out through the vale, keeping the winding road until I reached the brow of the orchard beyond, which was in full bloom.

The distant hills wore a delicate clear blue tone, and as I caught glimpses of them between the round hills about me, I distinguished Mount Æolus, that distant pile of Dorset marble far to the northeast of the Gap. Leaving the orchard, I crossed the road and entered the deep grasses of the old lake meadow, where the sphagnum is knee-deep. Here, as last June, the Indian Poke and cowslip blossoms freshened the borders of the stream. Along the edges of this wet region, I waded carefully until I reached the famous Spring of Arethusa, around the glacial hill to the left. I searched in the open meadow beyond the mill for Pitcher-Plant blossoms,—and found many in full bloom. The grasses were ablaze with tasselled sedges and nodding flowers of Iris,—a sight well worth a long journey to see.

I rounded about the swamp, and passed out at the north end, near Washon Bridge House. Here I ascended westward,—over the knob-like hill north of Pownal Pond. On the opposite slope I descended, finding nothing but trees and fences in my way. I observed a hollow-hearted chestnut tree,—a shell and nothing more. I could scarcely see where its green branches could gain nourishment. The leaves were, however, the largest in the wood, and the buds were perfect. The heart of this old tree was an empty, blackened space, the outer bark weather-worn and crumbling in decay.

Arriving on the north shore of the pond, I searched for the aquatic plant *Polygonum amphibium*, which I had observed last season along the muddy pools. The fencing of the sheep-pasture here debarred very free progress about the shore. I was forced to climb the hill for some distance to find an opening through the network of barbed wires. The day was warm, and the sheep had taken shelter in the shade of the pines on the hillside.

The small pine grove along the west shore of Pownal Pond is often used as a picnic ground. Years ago the south shore of this lake was clothed with dense oak, pine, and maple trees. These vales were the homes of many sturdy settlers while the fields were being cleared. The stone walls which they erected outlast the memory of their builders, and are the only monuments that time cannot remove. The few remaining gable-roofed houses with their gaping doors and windows, along the East Road, during the next few years will become obliterated entirely. The overgrown hedges of cherry-trees and grape-vines are still struggling for existence by the road; while the cinnamon-rose and southernwood are choked amid the cat-mint and burdock along the border of the door-yard path.

These vales of Etchowog are deserted, and the thrift of the Revolutionary days has departed. Nature is returning to her pristine state, and seeks to subdue these traces of man by covering all with weeds, slow decay, and mould.

Once in the pine grove, I discovered that I was in the vicinity of a small cabin, which stood on the brow of the hill overlooking the pond. A door opened southward from the house, and pasted upon it in bold handwriting was the declaration that it was inhabited.

"Rented by Edward Green, Esquire.
Do not trespass on these premises."

The Beautiful Arethusa. (*Arethusa bulbosa.*)

This is a rare, shy orchid found in company with the Rose Pogonia and the Grass-Pink in the heart of sphagnous swamps.

The water along the muddy edges of the pond displayed innumerable wriggling pollywogs and small fishes. About midway along the shore, I found the *Polygonum* in blossom. I recognized the pink clusters nodding on the water at some distance from the bank. The wind, blowing in little whirling gusts, ruffled the waves. The distant Yellow Lily pads (*Nymphæa advena*) flapped strangely for an instant or two,—turning their great round leaves over on the water's surface, and displaying their crimson linings.

I now devoted myself to solving the great problem of snaring the Lady's-Thumbs of this deep-water species of *Polygonum*. They were just beyond my reach, and I was obliged to drag up an old weather-worn, decaying pine, and float it out to walk upon. With a staff in one hand and a willowy snare in the other, I ventured out upon the bridge as far as I dared to go. I managed after many a slip to snare off the blossoms and float them in to shore. On June 26th I was able to secure some of the flowers of *Polygonum* growing in the centre of Thompson's Pond, and found the two plants identical.

There are seventy-one species of this genus in North America, and about two hundred reported for the world. The above species, found in our lakes and ponds, is not rare, yet it is seldom observed in clear water. It was for me a new discovery for this region.

I was pretty well soaked after wading around these muddy shores, and not a little tired with the planning and building of bridges. I rested, therefore, on the hillside among the ferns, watching the daring devil's-darning-needles—dragon-flies—come and go about my head. The name of darning-needle is still full of alarm to me, but the dragon-fly is harmless both in name and nature. Bees were busy humming at their duty, frogs were croaking the hours away, and the wind was still flapping the ancient pads of *Nymphæa*, while low, sweet tones through the forest crept. I could have fallen fast asleep here beneath these shades, yet I was far from home, and my boots were heavy and wet.

I made slow progress homeward to-day, with my heavy foot-gear and vasculum. I followed the dusty road to the Ball Farm gate. Here I turned into the old grassy way which had been in use before the present road was built near Thompson's Brook. One can scarcely trace a track of the traffic of the past years in the present sod. The stone walls on either side of the lane are hidden with woodbine and red-raspberry bushes. Beside this path towers a great pine tree. I had promised myself a long rest beneath this shade, and gladly threw down my pack, and made a pillow of my tin can.

The fleecy clouds rolled across the infinite blue over my head, and a sense of relaxation and solitude stole over me. I must have fallen asleep, and I was suddenly aroused by the cawing of crows that were circling above me,—wondering perhaps whether Major and I were in a proper condition for their approach.

I was more tired after my rest than before, and I began to question, as many of my neighbors had done, the wisdom and profit of my bog-trotting. Well, my neighbors see no value in pitcher-plants and sundew. They say there is no money in them, and pity me for investing my time as I do. Neither do I understand why the farmer chooses to cultivate squash rather than follow some other occupation. It is his business to cultivate squash as it is my business to cultivate sundew. Some crops are failures in their monetary returns,—others in their yield of pleasure. As many wish money only to procure pleasure, if pleasure can be procured without it, why not take the easy way? The end is the same without the worry of the squash-bugs, and the weeding and killing of the crop,—to say nothing of selling the fruit. The sundew plant would die were it to exchange its habitat for that of the squash.

Giving myself a shake, I arose and again started on my way. Once through the fence, I nailed fast the board I had loosened, and climbed up to the road through the blackberry briars.

I did not make another journey for a week or more. On June 10th, I ventured through the Glen of Comus to see the colony of the two hundred Moccasins. An albino—a pure white flower of *Cypripedium acaule*—was found recently by a lad in the district. He reports that he collected it amid a group of thirteen Pink Moccasin-Flowers, apparently the only pale one of the sisters.

Upon close examination of the structural parts of the albino, I observed that the left anther had not developed at all. It appeared blasted in embryo, and now looked like a brown smeared spot. The sepals and lateral petals were of a rich chrome yellow. The dainty labellum was pure white, of a pearl-like texture in the veining, and tinged with chrome on the crest of the moccasin. It was indeed a strange, beautiful flower.

I had always supposed that an albino of any species of orchid was pure white throughout its parts, and was therefore surprised to find the sepals and side petals yellow.

Albinos of this species have been collected in this district for four seasons. A colony, found near the schoolhouse, produced six white blossoms. The children, calling them faded Pink Moccasins, believed them to have lost their color after maturing. It appears from its persistence that the variety is permanent, and not the freak of a season. The abnormal anther may be present in all albinos. If so, it is evident that evolution is taking place in the Pink Moccasin-Flower through the suppression of one anther in genus *Cypripedium*, which possesses two, while all other genera of the family have but one anther.

The colony of the Showy Lady's Slipper in Rattlesnake Swamp, producing forty-two blossoms in 1899, unfolded but fifteen flowers this season. For reasons unknown to me, it was not a good year for Cypripediums.

Notes:

[1] Milton, *Lycidas*.

The Rattlesnake Plantain. (*Peramium.*)

A group of three species collected on Rattlesnake Ledge, Mount Œta, Pownal, Vermont.

12
Saucy Jays and Polypores

To arched walks of twllight groves,
And shadows brown, that Sylvan loves,
Of pine, or monumental oak,
Where the rude axe, with heaved stroke,
Was never heard the nymphs to daunt,
Or fright them from their hallowed haunt.
—Milton, *Il Penseroso.*

I followed down one of those sun-dried brook beds that melting snows from the hillsides had eroded during past ages. It proved a short journey to the Glen of Comus, descending northward toward Ball Brook in the vale below. I had not proceeded far when I discovered what at first sight seemed a robin's nest, built high in the branches of the American Hornbeam,—or, as it is locally known, the Iron-Wood tree (*Carpinus Caroliniana*). It is the only American species of this genus in the Birch Family. Several saplings stood about fifteen feet high, two having so interlaced their branches as to form a strong crotch about eight feet from the ground. The nest was fashioned roughly, built of small sticks, and fastened in the crotch-like loft of these trees. Looking more closely, I perceived the nest was a third larger than the robin's, and was not plastered with mud. I soon discovered that the bird upon the nest had blue tail feathers and a jaunty cadet top-knot, as she peeped over the edge of the nest at me. This was, then, the saucy jay's nest, so seldom found about these woods. She became disturbed, and flew off down the ravine. I managed to climb up the trees high enough to determine that the eggs were still unhatched.

The glen was dark here, and Major and I sat in the dim light beneath the shadows of this dense under-wood. The jays began in chorus to scream unmercifully. They were distressed by Major's presence, and flew saucily above his head. He scarcely knew what to make of it all,— not being a bird-dog,—and sat demurely looking at me and wagging his tail. Finally, tired of their own screaming, the jays proceeded down through the intricate windings of the hollow, and we heard their mutterings at a distance,—a pleasant wild sound through these forests.

I looked carefully over the iron-wood trees. They are not uncommon hereabout. Their trunks are ridged and muscular in appearance. These trees are in fact very strong, possessing the endurance of the oak and beech. They never attain great height,—from fifteen to forty feet or so,—but the weight of their wood to the cubic foot is forty to fifty pounds.

Many decaying logs of yellow birch and pine stumps were scattered along the brook bed. They were covered with beautiful mosses and fungi. The shelf-like growth, known as *Polypores*, was abundant on these trees. There are several varieties of this group of fungi here. The larger kind often attains a diameter from six inches to three or four feet, in a semicircle, according to age. It is a hard, leathery or cork-like growth full of pores, the top of the shelf seeming like a slanting roof, grained and striated as it were, with colored slates of gray and brown. This fungus seeks no special species of decayed tree, as I find it clinging to several,—the yellow and white birch, and hemlock logs and stumps.

The underside of Polypores is of a soft ashes-of-roses hue when fresh, later becoming a dull gray-brown. If one looks sharply at the under surface, even with the naked eye, he will observe little pores no larger than pin-points. Under the magnifying-glass, these appear like giant honeycomb cells. Cutting through a section of the shelf, we find that these pores penetrate the heart of the shelf. In these little pore-like cells, the spores or seeds are borne, more hidden even than those of the Fern Family.

The name *Polypores* originated from these minute pores. Puff-balls or toadstools spring up during a night in pastures or corners in rich wood. But the Polypores are slow in growth.

A beautiful species of the Polypores is worshipped by the natives in Guinea. I also have found and worshipped several specimens of great beauty. I discovered a very large shelf on a decaying hemlock stump in Rattlesnake Swamp, which I severed carefully with a woodsaw, removing enough of the stump to show its attachment to the tree.

As I passed through the glen to-day, I found many large and small specimens of this fungus, whose growth demands a humid atmosphere. The fact that decay does not take place rapidly save in a damp, warm wood, naturally proves that Polypores require such shades as these in which to develop.

Tall brakes rose luxuriantly four feet high or more. The atmosphere was heavy, and the sphagnum was steaming wherever the sunshine poured through the leaves upon it. A certain fragrance of the earth rose up from the swamp and met me everywhere,—a mingled perfume as of violets and Cypripediums. I explored about the pools to the left, finding many flowers in bloom.

Upon a miniature island in the centre of the pool grew the tall spikes of the Queen Moccasin-Flower, in bud. Turning to the south, under the hill among the rocks, is the fountain of the glen, which freshens the heart of the flowers beyond. Surely these are the haunts of thrushes, as well as the home of the queen of the orchids. The Golden Moccasin-Flowers peeped out from beneath the shades of ferns, and sprinkled the mellow glooms with jewels, like footsteps of sunshine left by the wood-nymphs of old.

The footprints of the woodman and the clips from his axe are yet unknown in this Glen of Comus. This is the sanctuary of the gods of old, and these the altars beneath the roofless temples, where man may worship still the deities of Nature. The wood-thrush's song rings through these cathedral aisles:

> "Untwisting all the chains that tie
> The hidden soul of harmony."[1]

I crept quietly through all these winding halls, which I had never before explored. Near the northern portal of the glen stood a white birch, branchless, and mellow in decay, yet beautifully robed with delicate Butterfly Polypores of a velvety purplish hue. Turning at the junction of the streams, I frightened up the oven-bird, the golden-crowned thrush. She moaned and fluttered away, as though in distress, dropping her wings and hiding among the ferns. I searched about for her nest, and soon found it low upon the ground. Her cottage door was open to the south, revealing five pinkish eggs mottled with purple. The nest was hooded,—thatched, as it were, like an Indian's wigwam, with leaves, twigs, ferns, and mosses,—so like the ground itself that I nearly walked upon it.

We have five true thrushes of genus *Turdus* in the Atlantic Region—the Veery, Wood-Thrush, and Hermit-Thrush are found in this immediate region. They are our peerless woodland songsters, coming about May 1st, and often lingering until September 15th. The Veery winters in Central America, and flies as far north as Newfoundland to nest in summer. Like the Hermit-Thrush, it builds its nest on the ground. The Veery has a mysterious strain likened often to an Æolian harp; the Wood-Thrush rings like the chimes of vesper bells, and the Hermit-Thrush has the deepest note of all, rolling like "anthems clear" through the dim woods. Burroughs translates its song thus: "O spheral, spheral! O holy, holy!"

My delight was complete, since I had found two rare birds' nests within an hour,—those of the melancholy songster and the screaming jay. Four days later I visited both of these nests to see the birdlings. The mother jay was not at home, so I did not distress her when I climbed up to peep at the homely babies. I passed on down to the deeper glen to the oven-bird's wigwam. She too was absent. Five little bald heads and five wide-gaping mouths were revealed as I drew near the nest, bespeaking the necessity of a thrifty mother to search for food to satisfy their needs. I touched their little heads, then drew back and waited almost an hour for the return of the mother bird, hoping to see the feeding of the young. But she was either shy or belated, and did not appear.

Notes:

[1] Milton, *L'Allegro*.

Third Season

13
The Swamps and Hills of Mosholu and Lowerre, New York

Within the circuit of this plodding life,
There enter moments of an azure hue.
—Thoureau, *Excursions.*

This season, on May 15th, I began my explorations in the hills and swamps of Mosholu and Lowerre. The most conspicuous flowers about these woods are trillium, spring beauties (*Claytonia Virginica*), bird's-foot violets, yellow violets, jack-in-the-pulpits, and pink azaleas. The swamps and slopes east of the Mosholu station are bright with these blossoms, which peep from the sod and shrub in their turn. In several places, also columbine, Dutchman's-breeches and dog's-tooth lily are abundant. During May these flowers, with the trees of snowy dogwood blossoms, fill the rolling hills and quiet valleys with delicate perfume and unrivaled glory.

Along the higher ridges, the brilliant Rock Pinks (*Phlox subulata*) bloom abundantly. Their mossy-mats creep over the hills from Bronx Park to Yonkers. They belong especially to the extreme southern part of the State of New York, and southward to Virginia and westward to Michigan. In these woods of Mosholu and Lowerre they flower immediately after the Dutchman's-breeches have faded. I had believed that these pinks must grow as far north as West Rock and the rocky heights of the Giant at Mount Carmel, as well as about the ridges bordering Lake Saltonstall, near New Haven, Connecticut. I was, however, disappointed to find that their territory extended no farther north than the wilder woods of New York City.

I discovered many beautiful plants of the Prickly Pear, or Indian Fig (*Opuntia Opuntia*) of the Cactus Family. It was named for a town in Greece where it grew. This strange relic of the primeval wood blooms in June, producing a sulphurous-yellow flower of great beauty. The large, spatulate-lobed, juicy leaves are sap-green in color, bearing many thorn-like spines. The new leaves, or lobes, appear as joints along the edges of the parent leaf. The fruit is edible. This species is often cultivated. It belongs natively to the rocky shores of Nantucket,

The Snowy Dogwood Blossoms, from the Hills of Mosholu, New York.

Like a drift of tardy snow,
Tangled where the trees are low,
Scented dogwood blossoms blow.
Dainty petals spreading wide,
Heart-shaped, lying side by side,
Not a leaf the flowers to hide.

Mary Wilson

Rhode Island, and to Manhattan Island. It is not abundant in Bronx Wood, however. Isolated colonies of the plant live in New York City, along the mutton-backed granite rocks in vacant lots, west of St. Nicholas Avenue, and along Washington Heights.

Wild Garlic, of the Lily Family, is ever present about the hills of Bronx Valley and Spuyten Duyvil Creek.

The Bird's-Foot Violet (*Viola pedata*) and the Round-Leaved Violet (*Viola rotundifolia*) seem to run riot on the Mosholu Hills, but it is not always easy to distinguish the species. A variety of Bird's-Foot Violet that grows here appears like a small pansy, and is designated as *Viola bicolor*, producing two delicate, velvety hues of blue-purple. The plant derived its common name from the shape of the leaves, which are divided into five to eleven pointed lobes.

The early Greek name for Violets and Pansies was *Ion*. According to Emperor Constantine, it arose from *Io*, a nymph loved of Jupiter. Nicander wrote that the name *Ion* was given to Violets because the Nymphs first presented Jupiter with these flowers in the fields of Ionia. They were known to Virgil as *Vaccinium*, and later in Latin as *Vittulæ, Violæ*, and to-day they are classified as *Viola*. Species of these plants were designated by the early Greek apothecaries as "*Herbes Bolbonac.*" In the sixteenth century plants of this family grew wild among the corn-field stubbles of England, according to Dodoens and Lyte. They were known as *Viola*, *Iacea*, *Herbe Clauellata*, *Pances*, Love-in-Idleness, and Heart's-Ease.

The Downy Yellow Violet (*Viola pubescens*), although not so common as blue violets in Bronx Woods, is abundant in special corners among the damp hillsides. Here, too, the Sweet White Violet (*Viola blanda*) dwells near the borders of streams. It is delicately fragrant, although not so sweet-scented as the Canada Violet (*Viola Canadensis*) growing northward as well as southward along mountainside streams. The perfume of the Canada Violet is much like that of the Small Yellow Moccasin-Flower.

Jack-in-the-Pulpit preaches from many rocky hills and hollows in Mosholu and Lowerre, where grow the largest plants I ever saw. They spring from bulbous turnip or onion-like roots, and are sometimes called Indian Turnips. These plants were known by Pliny in Christ's day as "Dragons," on account of the stalks, which are speckled like an adder's skin. The ancients believed that the leaves of Dragonworts, carried in the clothing, would prevent stings of vipers. Others believed that the leaves, wrapped around cheese, would keep it from mouldering.

Matthiolus thus described the Skunk-Cabbage of this group, to which was attributed mythical properties, since it grew sparingly in northern Asia: "Great large leaves, folded and lapped one within another, with an upright stalke, at the top a floure like to a spikie-eare."

The Green Dragonwort and Jack-in-the-Pulpit were known, until recently, generically as *Arum*. *Arisæma* antedates *Arum*, referring to the red-blotched stalks of some species. Jack-in-the-Pulpit to-day is known in the science as *Arisæma triphyllum*. The origin of the name Jack-in-the-Pulpit is recent, and, like Indian Turnip, is purely of American origin. Clara Smith of Medford, Massachusetts, so christened these Dragonworts, in a poem which was sent to Whittier for revision. He published it in *Child Life*, about 1884, after amending and adding several lines. The poem became popular, and the flower was thereafter known as "Jack-in-the-Pulpit."

Columbine was especially plentiful along the hillsides; and the hollows and crevices of rocks were filled with blade-like leaves, resembling Sweet Flag (*Acorus Calamus*). They proved, however, to be the leaves of the Blackberry Lily (*Gemmingia Chinensis*). This lily creeps from southern New York to Georgia. The seeds resemble blackberries. The plant produces several large blossoms in a terminal bracted cluster, of an orange color mottled with purple. This species was formerly known as *Pardanthus*, meaning a Leopard-Flower. The roots are of a golden color.

Returning from the Point of Rocks above Deer Park, I passed along lanes bordered with cedars and junipers, while violets, rosy-pinks and tufts of maidenhair spleenwort clung to the ledges. On leaving the swamp below, I found a drowsy diamond snake in a stupor, from having recently swallowed a bird or frog. The diamond-shaped checks upon his skin betrayed his species. He is considered venomous, therefore I remained a safe distance from him.

On May 18th I again visited the Point of Rocks and McLean's Woods, searching for *Orchis spectabilis* and for *Cypripedium parviflorum*. Leaving the car at Bedford Park, we struck westward, coming out near Poe Park, where still stands that quaint white cottage in which Poe wrote *The Raven*. We bore around the slopes, northward beyond the Racing Park, and entered a country lane, soon turning again to the left into the forest, where stood great pools of water. Along the sluggish stream grew many rare species of fern. Finally we entered Jerome Avenue, leading toward Yonkers.

We searched the borders of the roadside for that little two-leaved orchid, Twayblade (*Leptorchis liliifolia*), formerly known as Lily-Leaved Liparis, which grows here in the moist woods. We were too early for it, however. We turned off into the deeper woods till we came to the tangled edges of McLean's Swamp. Here, a little later, I collected pink azaleas and marsh marigolds, golden-ragwort,—known as the False Valerian (*Senecio aureus*),—white mustard, and watercress,—also of the Mustard Family. Throughout these beautiful woods the Broad-Beech Fern (*Phegopteris hexagonoptera*) and the Oak Fern (*Phegopteris Dryopteris*) dwell. The Jack-in-the-Pulpit and Indian Cucumbers were here, towering above the False Lilies-of-the-Valley and Trillium; and among these vines the dainty golden-shoes of the Fragrant Cypripedium tripped forth like fairy's foot-gear. The Indian Cucumber (*Medeola Virginiana*) is a strange plant belonging to the

1. Indian Pipes. (*Monotropa uniflora.*) 2. Pine-Sap. (*Hypopitys hypopitys*).

Humbly it wears its robe of snow,
When summer gives its bud release,
And Indians called it long ago
The Calumet or Pipe of Peace.

W. M. L.

Lily-of-the-Valley Family. The root is tuberous, of a white, brittle substance, with a flavor like that of the garden cucumber. The leaves occur in two whorls along the slender stalk. About the middle of the stem there is a whorl of five to nine oblong-lanceolate leaves; above this another smaller whorl occurs, with three to five leaves; and still above this, all the small flowers nod toward the ground.

Medeola is an adaptation of Medea, the name of the daughter of King Æetes of Colchis, who aided Jason by her witchcraft, and was afterward deserted by him. This plant is supposed to possess magic properties as a medicine, and is thus used by the Indians and other superstitious folk.

Colonies of Brake and Clayton's Fern grew in the hollows on the hill, and about the stagnant pools northward. We wandered up and down the slopes with eyes bent upon the ground, carefully pacing between the vines, searching for the Golden Moccasin-Flowers. Most of the buds still were folded within the sepals, although we found a few that were showing their golden tips and carmine lacing-petals. In the swamps beyond, we gathered a dozen Nodding Wake-Robins (*Trillium cernuum*). This species is not so gaudy as the Painted Trillium, being of a delicate rose-pink, and often pure white.

Later in the week, while exploring to the east beyond the lane toward Mount Vernon, I ran upon a select group of *Cypripedium parviflorum*, many stems of which bore two blossoms. This colony grew beneath pines, cedars, thorns, and dogwood trees. The soil was rich and dry, and the leaves, for the most part, were blown off the bare black soil. I counted a hundred plants—evidently seedlings—besides the ones in bloom. Some were at least three or four years old; others were of only one or two seasons' growth. The smallest plants were but a few months old. I had never found so many natural seedlings of *Cypripedium* before.

We journeyed homeward through Putnam Valley to Mosholu, passing Van Courtlandt Mansion. Near Cold Spring, along the borders of the Golf Links, we found the handsome leaves of Bloodroot (*Sanguinaria Canadensis*), of the Poppy Family, which is one of the early flowers, and is pure white. The roots contain a reddish orange juice which looks like blood, whence the name is derived.

At the crossing of the Putnam Railroad, we passed over the bridge near Deer Park, east of Mosholu. Leaving the road abruptly, we turned to the right, following along the west shore of the Putnam Swamp, which is filled with the rank growth of skunk's-cabbage, Indian poke, tangled grape-vines, mints, mustards, golden ragwort, violets, dog's-tooth lilies, and unknown measures of wild ginger root, stick-tights, or "pitchforks," and cockle-burrs.

The Yellow Cypripedium also, at one time, grew along the edges of the swamp, amid the Indian poke, violets, and lilies. Slowly we climbed the hill toward the northwest, along winding paths among white birch saplings, pines, and junipers, until we reached the Point of Rock near Lowerre, this pile of granite being the highest along the Putnam Division, in this vale. On the east of the rocks, we found a dozen or more plants of the Showy Orchis, scattered among the stones and vines. Their flowers, however, were faded, and fell as soon as touched. Among the low bushes and plants I found a colony of the leafless parasitic Cancer-Root (*Thalesia uniflora*), of the Broom-Rape Family.

Another plant lacking green leaves is the Indian Pipe or Corpse-Plant (*Monotropa uniflora*), frequently met among the Orange Mountains of New Jersey, and throughout the Hoosac Highlands. It grows among decayed brush-heaps in dark woods, during June and August. There are twelve species of *Monotropaceæ*. The flowers of a sister genus of *Monotropa*, found in the Sierra Nevada Mountains, are remarkable for their deep rose-carmine coloring. It is sometimes known as the Snow-Plant (*Sarcodes sanguinea*). The tall club-like, fleshy spikes are encircled with crimson pipe-shaped flowers, often sixty or even a hundred being found on one spike. A specimen plant, collected in Washoe Valley near Franktown, Nevada, was sent to me last season, on May 15th. The flowers grow there along the higher slopes of the mountains, at an elevation from six to nine thousand feet above sea level, amid brakes, pine, fir, cedar, golden-chestnut bushes, and the beautiful evergreen shrub, Manzanita (*Arctostaphylos punges*).

Above the Point of Rocks, a rough canvas tent was pitched in a crevice of the ledge—probably the temporary abode of Italian green venders. Columbines, rock-pinks, violets, and Wood Betony (*Pedicularis Canadensis*) crept along our path. The plants of the Wood Betony produce yellowish-green as well as crimson-purple flowers. They are often called Lousewort. Children often misname them orchids, believing this term suitable to all odd-looking flowers.

The white Monumental Rock, east of Mosholu Swamp, is covered with glacial scratches. From its summit a dream of visual delight spreads toward the blue walls of the Palisades along the Hudson. The blossoming wood, waving with pink and white dogwood branches, the western slope of the rock itself, banked with rosy pinks, nodding lily-bells, and columbine, form a scene the impression of which never can fade from the mind.

Passing the station of Mosholu, we followed the path along the railroad southwardly near the marsh. Here, during July, three rare species of *Habenaria* will bloom. And in the meadows farther northward, the Ladies' Tresses—genus *Gyrostachys* of the Orchid Family—and the Blue Fringed Gentians (*Gentiana crinita*) will unfold in September and October. Soon we came to the end of the swampy path, and entered the broad meadows of Van Courtlandt Farm. In the distance the antique Colonial Mansion of Revolutionary fame stood out among the ancient trees. Over these fields the first bobolinks of the season were carolling. I found one of their eggs among the grasses.

The Snow-Plant of the Sierra Nevada Mountains. (*Sarcodes sanguinea.*)

It is a risky nesting-ground for birds. The parades of the militia form upon these fields, the regiments' camp being located east of the mansion. The trees along the lanes hereabout are English species, planted years ago by the owner of the mansion. Many are crumbling and decaying as the mills to the left. Another half century will do away with such as these. To the right flows Spuyten Duyvil Creek,—a small, elusive stream; and as it glides into the swamp beyond, it covers much marshland, where sedges and cat-tails flourish and no one dares to wade.

14
The Swamp of Oracles—Hoosac Valley

> Science is welcome to the deepest recesses of the forest, for there too nature obeys the same old civil laws. The little red bug on the stump of a pine,—for it the wind shifts and the sun breaks through the clouds.—Thoreau, *Week on the Concord and Merrimack Rivers*.

On June 6th I departed from New York for the Hoosac Valley, to obtain photographs of my orchids and their haunts. Rosy-faces, golden-slippers, witches'-bells, and milky-white stars all arose from the earth at once in gay array, and disputed their line of order in posing for their pictures. I had sent no forewarning of my coming to the swamps. I fancied I should find more flowers in bloom if I took them by surprise.

The morning of the seventh, I hurried off at sunrise through the dewy meadows. I felt sure I was too late for the Showy Orchis and the Ram's-Head Cypripedium, the former having faded in Bronx Valley as early as May 28th. The hills were glorious; the robins, orioles, and bobolinks were carolling joyously. The meadows, still heavy with dew, caused me to choose my path along the edges of the Bone Lot near the old Pond Hole. This I found fringed with pink azaleas,—the swamp-apple blossoms loved of the children hereabout. I entered the chestnut wood beyond, and sought the colony of the Large Yellow Moccasin-Flowers, only to find that the shoes had been broken from their stems, and that there were none remaining. Still, there were other groups in the Swamp of Oracles, and I proceeded to scout the slopes leading to the hollow below, winding about the knolls—or Sugar-loaves, as they are called here. These glacial hills are worthless barren pastures at best, seldom ploughed for rye or corn, for all attempted crops of grain here have proven thin and dwarfed, and when it rains gutters burrow in the hillsides.

As I descended through Patterson's Meadows, the air was musical with humming bees and birds. Moths and butterflies sailed lazily about the pools below, hovering about the first blossoms of Fleur-de-lis. Over the rolling fields near, the tender leaves of Indian-corn rustled musically in the breeze, and crows were still lingering on the fence, not in the least frightened by the snares and scarecrows about the field. I found the meadow ablaze with late Columbine (*Aquilegia Canadensis*); I had never before seen fields so luxuriant with the blossoms of this plant. They danced among the daisies, and outnumbered the grasses in their patches. The generic name *Aquilegia*, or *Aquileia*, is said in our manuals to be derived from *Aquila*, an eagle, since the curves of the hollow spurs of these flowers resemble an eagle's talons. But in this case, the name should read *Pes Aquilegia*. Among the ancient herbals, however, there is no record of this derivation. Originally, as Dodoens wrote in 1578, this plant was known as *Aquileia*. Aquileia was also the name of a town in the vicinity of ancient Troy. The town was celebrated in history for its desperate resistance to Attila, King of the Huns. Assuming that the origin of the name is vague, and observing the customs of the ancients in the naming of plants, it might be inferred that these flowers were first observed in the town of Aquileia, or were named in honor of a king or herbalist of the region. This was the case with the *Pæonia*, which took its title from that good old man, Dr. Peon of Pæonia, in Macedonia. The origin of the common name Columbine, also, has occasioned of late much discussion in the popular plant journals.[1] One author, claiming that the spurs of these flowers resembled a dove's-foot, said that the name should read "*Pes Columbinus*" —*pes* meaning foot, and *columbinus* signifying dove. But "*Pes Columbinum*" was used by the ancients to designate an allied group of *Aquilegia*, a species of wild geranium, written of by Linnaeus in 1753 as *Geranium Columbinum*. It is commonly known in English as Dove's-foot Geranium, and in French as *Pied de pigeon*.

According to Gaza, species of *Aquilegia* were supposed to have been named originally by Theophrastus—centuries before Christ—*Ponthos Theophrasti*. Theophrastus is accepted as our first real botanist leaving extant records of plants. The name *Columbine* for these species appears to have originated in England, or in the Low Countries. Dodoens described them under that title as early as 1578; and as botany was not actually revived until 1530 and 1542, we may accept Dodoens as authoritative on the common names of that day. He writes of the Columbine: "The shape and proportion of the leaves of the floures do seem to represent the figure of a dove or culver,—these floures produce hollows with a long-crooked tayle like a Lark's-claw (and bending somewhat toward the proportion of the necke of a Culver)."[2] The honeyed-spurs of Columbine, therefore, suggested the curve of a dove's-neck rather than the dove's-foot or the eagle's-talons. Another author is reminded of a "dove's-cote," as he looks into the open flower, which seems to him a fitting home for doves.

Columbine-flowers are often called "Honeysuckles" by children. The name Honeysuckle, however, applies to the Woodbines which

Dodoens describes as growing with flowers "in tufts like nosegaies, of a pleasant color, and long and hollow almost like the little bags of Columbine." The Columbine became confused with the Honeysuckles of the Woodbine Family, since Columbines produce "little bags of honey"—which the children sucked and christened. Furthermore there is a resemblance in the long hollow spurs of the Woodbine flowers to those of Columbine blossoms.

The Columbines belong to the Crowfoot Family (*Ranunculaceæ*), and are closely allied with sister genera, including Clematis, Anemone, Hepatica, Meadow Rue, False Bugbane, Buttercup, Marsh Marigold, Goldthread, Larkspur, Aconite, and Monkshood. These species produce plants with cut leaves, as it were, resembling feet, claws, or talons of various birds, animals, and fishes.

Continuing my journey, I crossed the edges of Rabbit Plain, observing the low blue huckleberry bushes, laden with green fruit, and the flaming flowers of the deep pink azaleas. Through the bushes peered the white schoolhouse of District Fourteen. I wandered along the border of the wood just out of sight of the curious gaze of the children. A cow-path led windingly along the shades for a quarter of a mile. Near the bars above the Swamp of Oracles, I found a spike of the Small Round-Leaved Orchis (*Habenaria Hookeriana*) in bud. I blazed a tree above it, marking the spot for another day when the flowers should be in blossom. Crossing the East Pownal road, I turned into a hollow to the west, following along over decaying logs and pine brush-heaps. The ground sent up a rich pitch-like perfume as the sun poured down upon the mossy sod. Wild lilies were abundant here, producing the largest leaves I ever saw. Solomon's Seal, arbutus, and wintergreen leaves (*Gaultheria*) were creeping everywhere near the edges of the deeper wood. Within the denser shades, growing among pine logs and heaps of leaves, I found the Great Round-Leaved Orchis, so seldom found in the lower vales. It proved to be a seedling, too young to bloom. The leaves were like large saucers, and of a beautiful silvery green underneath. The plant is always suggestive of the luxuriant tropics. I marked the corner, and shielded it from any chance vandal eye with a broken branch of black birch.

The slopes leading to Cold Spring, in the hollow below, were abrupt, and I was forced to slide most of the distance, clinging to the bushes. I came out at the foot of the hill in the midst of a colony of Sweet Canada Violets (*Viola Canadensis*) in full bloom. They grew along the borders of a little brook flowing through a dense thicket of soft maple and black birches. I had never before found this species in flower here. It seemed to have flown down from the heights of the Dome, to grace this swamp. Belated purple birthroot and its sister, the painted trillium, were still nodding here. There were also a few pale-faced priests-in-the-pulpit, unlike the larger coarse purple ones found in Bronx Valley. These Indian Turnips are not abundant here as in the swamps and hills of Mosholu. The wild leek of genus *Allium* seeks the higher mountainsides.

I followed the Canadensis Brook to the edge of the Swamp of Oracles, crossing Ball Brook at the junction of these streams. I penetrated where the rarer orchids dwell, and where few children dare to travel. I was still too early for the Showy Queen Moccasin-Flower, but on time for the large and the small golden slippers, as well as the Pink Acaule—that humble two-leaved Cypripedium which, as a rule, only seeks the dryer edges of the swamps. The Large Yellow Moccasin-Flowers were beginning to fade and turn brown. The swamp was luxuriant in its growth of ferns and vines and foliage. Dogwood trees are very scarce here, but the azaleas, mountain laurel, or calico-bush, and the lambkill flowers make up for the missing snowy blossoms.

In the heart of the swamp I was attracted by an uprooted tree, about whose stump stagnant water had settled, now reflecting the shadows and sunshine as a miniature lake. Several baby deer-mice were in the pool. Many were dead, and the live ones were swimming about in desperation. I counted six or seven in all. I fished them out, and placed them on the sun-dried moss, which covered the roots of the turnover, forming little islands in the lake. But these white-faced, pink-eyed little creatures were no safer after my rescue than before; for soon, in their nervous fright, they crawled off the mossy islands, and were still swimming when—not wishing to witness the end—I went away. It was one of the many mid-forest tragedies which Nature seems to plan with so little philosophy, and which I knew I could not prevent. Had I removed them from the water again and placed them at a distance from the mud-hole over which they were born, certain starvation would have awaited them. In the topmost parts of the overturned stump, amid the roots and peat, a pile of forest leaves was rudely huddled, forming the deer-mouse mansion, hidden from the crawling turtles and creeping snakes, as well as from the hawks and owls in the trees above. There are many natural causes of destruction for such animals in the woods. Usually I have found the deer-mouse's nest in low thorn-apple bushes, at least six feet above the ground, but always near the borders of streams. Such nests at first remind one of a last year's bird-nest filled with drifted autumn leaves, until the little wild-wood albinos are discovered.

With my vasculum packed full of perfect blossoms, I started homeward, following the Pownal Centre road westward, in order to have a look at the Ram's-Head Cypripedium. On the edge of the marsh, as usual I found the two hundred unfolding buds of the Pink Moccasins (*Cypripedium acaule*).

Near the Amidon Meadows, I startled up two mother partridges and their broods—the Ruffled Grouse (*Bonasa umbellus*), so prolific in these woods. The old hens, fluttering and sputtering, limped away with their wings drooping, and continued to warn their chickens to hide. The little speckled fellows were soon lost sight of beneath the dead leaves at my feet. They ceased to peep, and being of the colors of the leaves, I hardly dared to advance for fear of stepping upon them. I sat down upon a stone by a tree, and waited for the return of the wild hens. Before long, I heard a rustling of leaves in the distance, and a clucking and calling as of a tame hen summoning her chickens to feed upon a worm. The little brown balls began—one, three, then a dozen, all at once to take their heads from under the leaves, and they ran like streaks of lightning. The mother partridge came so near, unawares, that I saw the color of her eyes. Finally, discovering me, she in terror signalled again, much as the tame hen does in real or fancied danger. The little grouse hid again, some of them putting their heads under leaves, while the body was wholly exposed.

On June 8th I visited Rattlesnake Swamp. Pink Moccasin-Flowers and late blossoms of Painted Trillium were abundant under the hemlocks along the slopes of the Domelet.

On June 10th I heard of a colony of albinos or white *Cypripedium acaule* reported on the Rabbit Plain north of the Swamp of Oracles. In searching unsuccessfully for it, I frightened up an old mother whippoorwill. She feigned broken wings, attempting to distract my attention from her two unprotected yellow eggs upon the leaves at my feet. Both partridges and whippoorwills remain on their nests until almost stepped upon, as a rule, believing that they are concealed because of their dead-leaf ground-coloring. The old whippoorwill perched on some distant pine logs, and moaned piteously while I looked at her eggs. Her great round, sad eyes distressed me, while she gave forth a sighing sound. I broke down a small tree over the nest and near the path as I left, hung my linen collar on a tree, marking the line of entrance for another day.

Four days later I returned, and found two little round balls of yellow down, just out of their shells which were lying near. Creeping up softly within touch of the mother, I had a chance of observing her carefully. She had no shelter or protection but a leaf of the False Lily-of-the-Valley (*Unifolium Canadense*) which covered her eyes and part of her head. She never stirred a feather nor blinked either of her round brown eyes. Close to the earth like the leaves themselves, pressed down with winter snows, it was difficult to distinguish her feathers from them. I finally frightened her from the spot. The poor little birds heard their mother's cry of alarm, and, babes as they were, instinctively understood it all, opening their dreamy sad eyes, and trying to hide away. Nest they had none, and rolled about over the leaves. I visited these birdlings so often, in my eagerness to make observations, that the mother finally left her young. One cold night, finding them almost freezing and starving, I took them home. They did not live more than a week, however, on account of my ignorance as to what food to give them. During this time they became very tame and dependent upon my care, rejoicing strangely when I came near.

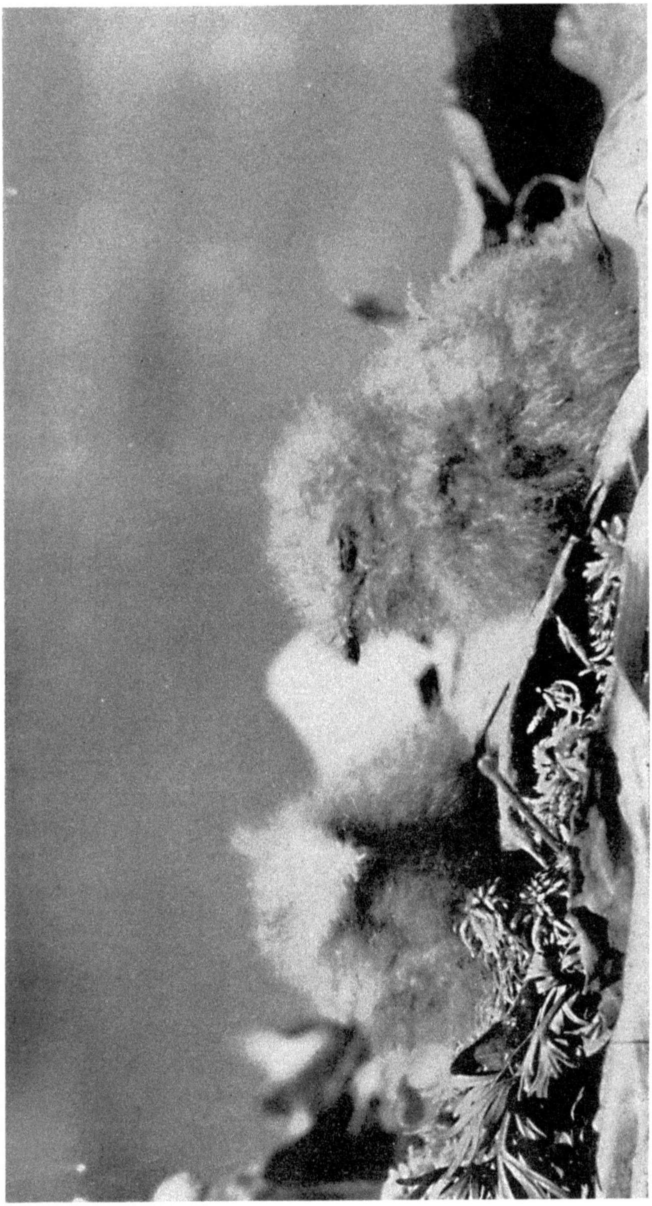

Motherless Baby Whippoorwills.

The Southern Chuck-Will's-Widow, a species closely allied to our Whippoorwill, builds no nest, but is said to move her eggs and young, in her large mouth, from place to place, wherever she may choose to abide. It would be well if Nature had thus taught our Northern Whippoorwills.

I continued to visit the Bogs of Etchowog, collecting azalea, iris, and the other flowers in their turn. In circling the Pownal Pond one day, I ran upon a Water Thrush (*Selurus noveboracensis*) and her brood of five little foolish, half-grown thrushlings. The awkward birds ran peeping across my path, not in the least afraid. I caught them all, and placed three in my hat, leaving two for consolation to the mother, while I hurried home to obtain a photograph of my prizes. But I was not able to reconcile them to their new conditions and food so easily as I had domesticated my whippoorwills. As soon as I had secured a negative, I returned them, nearly famished, to the mother, who was running along the shore of the pond, tipping-up her tail like the wagtail. These birds are swift in flight, skimming near the water, whistling as it were, while they catch insects. Their nest is very difficult to find, being as a rule among the roots of trees along the shores of ponds or streams in damp woods. I frequently observe these birds walking in the stony brook flowing down from Cold Spring in Chalk Pond region, as well as about the shores of Aurora's Lake in North Adams.

The hillside clearings in this region are the haunts of woodchucks (*Arctomys monax*). Many holes show where they have burrowed. Usually these ground-pigs seek for their habitations clover and bean fields, which furnish them provender. Exploring the door-yard of the woodchuck, I found several plants of the Small Round-Leaved Orchis maturing their seed-capsules. Not every wild pig's garden bears this evidence of aestheticism.

The fertilization of these strange Round-Leaved Habenarias is unique. The anther is eager to give up its pollinia. The adhesive masses shot from their cells when I touched them, and fastened to the head of my hat-pin. When placed near the viscid surface of the stigma, they were drawn forcefully from it, thus impregnating the ovules in the ovary. These masses of pollinia, once glued upon the thigh of an insect, would remain there until deposited on the attractive stigma of their proper species.

On my next excursion to the Bogs of Etchowog, I found nothing new, save six spikes of the Small Purple-Fringed Orchis in bud. I was too early for Pogonias and Limodorums, which are fast disappearing from this swamp. The colony of Fragrant Yellow Moccasin-Flowers, in the Glen of Comus, was photographed one morning while the sunshine struggled in through the leaves, lighting up the flowers in this labyrinth of tropical foliage. They were fragrant in the highest degree—a true form of *Cypripedium parviflorum*, with a slight variegated effect of carmine coloring on the tips of the slippers. This is the first

A Colony of the Small Yellow Fragrant Moccasin-Flower (*Cypripedium parviflorum*) in the Glen of Comus, District Fourteen, Pownal, Vermont.

There's a haunt I would lead you to,
Home of the gossamer and the dew.
Where, from out of the murky loam,
Springs the sacred flower of the gnome.

Clinton Scollard

instance of such spots of crimson on the exterior observed by me. Near this group stands also a larger colony of *Cypripedium hirsutum* seedling plants. Several had bloomed this season. One slipper had been destroyed by what appeared to be a snail. Nothing of the flower remained but the column, with the adhering anthers and stigmatic lobes. The sepals and petals, including the labellum, were eaten away. The snail was still clinging to the column, and must have found some delicate food in the juices of such golden petals to cause him to tear the flower apart He may, however, have fertilized the species in the act: yet the destruction of its parts would have weakened the possible chances of the seed-cape maturing properly.

In the bend of Ball Brook, amid the ferns, the Tall Northern Green Orchis (*Habenaria hyberborea*) blooms, its seeds having floated down here from the seed-capsules ripening on plants bordering the stream above.

Wild Ginger-Root (*Asarum Canadense*) grows abundantly along the sphagnous edges of the Swamp of Oracles. This plant produces bell-shaped blossoms, which invariably turn downward, hiding in the soft soil beneath the leaves. Its creeping roots are of a spicy, ginger-like flavor. The leaves—kidney-shaped—appear as small burdocks at a distance. The generic name is very obscure, although the plant was known to the early Greeks, and later known in Latin as *Asarum*, *Nardus rustica*, and *Perpensa*. Macer called it *Vulgago*, while it was known in England and Germany in 1578 as *Asarabacca*, *Folefool*, and *Hazlewort*. It was used by the ancients as an antidote for venomous serpent bites, sciatica, difficult respiration, and various other diseases.

On June 15th I made my farewell journey to Etchowog. Turning into the thicket, east of the Barber Mill, I followed a path as far as possible, and then waded through sphagnum into a meadow-like clearing of three acres or more, concealed in the deepest of solitude. It was closed in on all sides by low alders, willows, and beautiful green spires of tamarack-trees. The sphagnum was many feet deep, spangled with flowers; and rising above the swamp grasses were iris blades and buckbean leaves. It was a little world whose limitations were the infinite blues above, the depths of moss below, and the circling green-fringed forest trees. The sunshine knew the field, and poured in upon it. I was obliged to wade slowly over the quaking sphagnum, assisted by pine-slabs, strewn about as stepping blocks.

The oblong green leaves of the rare Buckbeans (*Menyanthes trifoliata*)—found also in the Cranberry Bogs, north of Pownal Pond—were here thickly entangled over the greater area of the meadow. A few spikes still were in blossom, although the greater portion were adorned with the bullet-like glossy, smooth seed-pods. Later in the season they would slowly ripen, and throw thousands of seeds broadcast over the sphagnous field. It is evident that this plant—so infrequent in its general distribution—is most productive of its own seeds in its chosen haunts. This species is a sister genus of the Blue-Fringed Gentians, abundant along the edges of these bogs during October. Gentians derived their generic name from King Gentius of Illyria, who first used them in medicine.

The Floating-Heart (*Limnanthemum lacunosum*), closely allied to the Buckbeans, grows also in our marshy pools, the leaves being heart-shaped, instead of oblong as those of *Menyanthes*.

In the middle of this swamp an island arose, over which grew willow bushes and tamarack spires, interspersed by grape-vines. I crawled through the bushes without finding a flower worthy of description. Surrounding the edges of this island, tall spikes of the Fragrant Northern Orchis (*Habenaria dilatata*) rose above the water-soaked sphagnum. I was able to reach a few of them, then sought the *terra firma* of the tangled swamp beyond. I ran great risk, since I was forced to wade the soaking bogs where the cat-tail flags were dense. I managed to jump from hummock to hummock, not waiting for the grass to grow beneath my feet. Beyond I struggled through the low tangled trees covered with the Wild Frost Grape-Vines or Possum-Grape (*Vitis cordifolia*), amid tamaracks, swamp-maples, poison-sumach and ivy-vines. I observed many enormous colonies of Pitcher Plants, still in bloom in the shades. Finally I reached the muddy bank of Ball Brook, ragged, dirty, and tired. I found the stream impassable because of the mud. Even old Major had sense enough not to go too near the stream. I was forced to make my way, as well as possible, back to the mill, among piles of old tinware that had been accumulating since the early Revolutionary days of 1777.

Once out of this place, it was a pleasure to enter the open Pitcher Plant Meadow, where I searched for Pogonias and Limodorums without success. I circled about the swamp and turned away from it at the north, climbing over the hill above the Washon Bridge, toward Cranberry Swamp. Blue jays were screaming loudly, and catbirds were mewing in the bushes near the pools. I found the path near the pond, which led through luxuriant ferns to the shades of pines beyond. Here the ground was carpeted with fragrant needles and cones. Bullfrogs croaked hoarsely in the swamp beneath the lily-pads, and over the hillside crept yards of the evergreens known as Ground-Pine (*Lycopodium obscurum*) and Club-Moss (*Lycopodium Selago*), known to the Greeks as Wolf's-Claw. This moss takes hold of the earth with its small roots, like the claws of a wolf.

This corner of Etchowog was the home of the mosquito, and I was obliged to use a bough of sweet-fern to keep the pests from devouring me.

"Fair insect! that, with threadlike legs spread out,
 And blood-extracting bill, and filmy wing,

Does murmur, as thou slowly sail'st about,
....
Thou 'rt welcome to the town—but why come here?"[3]

Notes:

[1] *The Plant World*, July, 1900; February, 1901; September, 1902; November, 1902.

[2] Dr. Rembert Dodoens, *History of Plants*, Lyte's Trans., 1st ed., p. 119. 1578.

[3] Bryant, *To a Mosquito*.

15
White Oaks and Gregor Rocks

> I can recall to mind the stillest summer hours, in which the grasshopper sings over the mulleins, and there is a valor in that time the bare memory of which is armor that can laugh at any blow of fortune.—Thoreau, *Week on the Concord and Merrimack Rivers.*

I had been on the trail for white Moccasin-Flowers for years; and on June 16th a lad of White Oaks Valley promised to guide me to the Forks of Broad Brook, and show me a colony of absolutely White Lady's Slippers. We arrived at the junction of the Field Brook—where it crosses the White Oaks Road near Richmond's Farm—and turned our horse's head through the fields eastward along the rude loggers' path travelled in winter. We were obliged to cross fields of oats and potato vines in order to arrive on the summit of these rounded hills. Here, amid the white birches and sweet-fern bushes, we fastened our horse. Among these ferns and briars I discovered five enormous orange-yellow mushrooms, which, apparently, were of recent growth. They were gorgeous to behold, and smelled like new-made bread, yet they were extremely poisonous. They were, upon examination, found to spring from a socket, above which a ring encircled the stalk. This is characteristic of genus *Amanita* of this form of fungus. This poisonous species with some susceptible people produces serious results if only handled, or if its fragrance is inhaled. I collected three specimens, however, and put them in my vasculum.

We now descended the slopes eastward leading to the Wilsey Lot, where we found a road leading up through Broad Brook Valley to the Forks. The path was bordered with tall, luxuriant brakes at least four feet high. They were covered with dew, and brushing against them, we became wet through. My guide was an alert observer, and darting off here and there into the ravine, he brought forth gay blossoms of the Showy Queen of the Moccasin-Flowers. As we proceeded, we came to a bend of the brook and followed along high ledges of rock, where we crossed to the right over the boulder-filled stream. A quarter of a

The Mountain Laurel. (*Kalmia latifolia.*)

And all the rugged mountainside
Thro' billowy curves is seen;
The roadsides meet in ample shade,
With showers of light and golden glooms.
And bubbling up the rocky ways
The clustered laurel blooms.

Elaine Goodale

mile more brought us to another bend in the brook, and here we re-crossed, and at the left hand abruptly climbed the hillside in the sphagnous bed of a rivulet. Here, my guide said, were the Pure White Moccasin-Flowers. They proved to be pale pink blossoms of the Showy Regina, however, and not, as I had hoped, the rarer *Cypripedium candidum*, or even the albinos of *Cypripedium reginæ*.

It is said that in a swamp near the Forks both of the Yellow Moccasin-Flowers bloom. American Mountain Laurel, the beautiful Calico-bush, was in full bloom hereabout, so the day was not without some new treasure found.

The wildness of Broad Brook Valley is delightful. The stream rises in luxuriant swamps on the eastern summit of the Dome, between Stamford and Haystack Mountains. The Forks along the stream are formed by this one and others flowing down from Mount Hazen, which lies to the southeast of the Dome. The valley is comparatively wide, and the stream, as its name implies, is broad. The chasm bears scars of days when the heights to the northeast were capped with glaciers, towering thousands of feet above the present mutton-backed summits, which were formed into their dome-like shapes by the erosions of this ice sheet. The channel of the stream is full of tumbling boulders, and during April, when the snows are melting, a wilder brook is unknown in the Hoosac Valley. Three seasons it has become so rough and swollen that it has carried bridges and all else in its course before it, threatening the houses and little chapel, as it rushed downward to the Hoosac.

Bears inhabit these dark ravines, and wander close to the habitation of man in the Hollow. Not far from where we collected our flowers, a bear had been killed last season by two lads fishing in the stream. As I left the glen, and drove out over the moss-grown hills, and through the hollows, I found the ground red with wild strawberries. Needless to say, I paused until I had my fill of this luscious fruit, and I carried a birch-bark cornucopia of it away with me.

On June 18th I visited my great colony of Showy Regina in Rattlesnake Swamp. As they were not yet unfolded in their perfection of magenta coloring, I put up a warning on a tree near them not to rob the colony until photographed,—fearing some fisherman would behold and gather the blossoms. However, they were photographed successfully on the 20th.

The stumps and trees in this corner of the swamp are covered with dead boughs, laden with lichens and reindeer moss. A kind of moss known as *Usnea* hangs from the boughs of the trees above. The whole region is humid and luxuriant, and could almost deceive one into believing that he was in the jungles of the Southlands, instead of among the glooms of the Green Mountains. The beautiful Butterfly-fungus (*Polypores*) is especially interesting throughout this swamp, growing on dead trees and logs. Another pretty species, found on stumps and the earth, has scarlet-tinted cups, nestling in early spring amid the mosses. The trees and stones display both their gray and foliaceous lichens everywhere hereabout; and in the fields, the smoking puff-balls burst beneath the footsteps. Foxes Fire-Eyes is common in this region. It is decaying wood, green in color, said to be full of threads of phosphorescent fungi. During the night this wood gives out a soft, luminous light, which if it happens to come from a large stump, often frightens both travellers and horses along our woodland roads.

In the Swamp of Rattlesnake Brook may be found the Pitch or Torch-Pine (*Pinus rigida*), shad-bushes, white and black birches, chestnuts, high huckleberries, and small bushes of the Ague Tree (*Sassafras*), which seem rare here, but are abundant in southern New York. The odd spires of the double or black spruce are also found among the denizens of this region. From May until late November the swamp brings forth, in their season, arbutus, mountain snowberry vines, St. John's wort, low huckleberry, the evergreen leaves of *Gaultheria*, prince's pine, creeping evergreens, numerous rushes and sedges. Here, too, the goldthread entangles the roots of mosses and trilliums, while the Sundew (*Drosera rotundifolia*), creeps along the mossy sides of the wood road, and in the deeper sphagnum about the stream. The rare Large Whorled Pogonia (*Pogonia verticillata*), of the Orchid Family, has been collected in this swamp for three seasons. This orchid is rare in New England, save in Massachusetts and Connecticut. It was first found in Vermont near High Bridge and Colchester by Messrs. Robbins and Oakes, the pioneer botanists, who passed through the State in 1829. The delicate emerald green leaves of *Clintonia*, marsh marigolds, Solomon's seal, Shin-Leaf (*Pyrola rotundifolia*), liverwort, wild briar-roses, lambkill, blue lobelias, Labrador tea, yellow loosestrife, blue-fringed gentians, innumerable ferns, the spikes of the Tall-Green Orchis, plants of the Round-Leaved Orchis, the Pink Moccasin-Flower, and rarely the beautiful orchid, *Arethusa bulbosa*—all of these conspire to make the region a wilderness of beauty.

On an excursion to Thompson's Brook, June 19th, near Meyers's Sugar-Bush, I collected ferns and iris. As I descended to the hemlocks, near the waterfalls, I stumbled upon the late plants of the Showy Orchis (*Orchis spectabilis*), in bloom—which were fully two weeks past the regular flowering date. They had faded in the hills of Mosholu on May 19th.

I had heard of Wash-Tub Brook for years, and on July 5th started off to explore the valley and the cliffs of Gregor Rocks above North Pownal. A limerock ridge runs from the base of Mount Anthony southeasterly to the Glebe in Witch Hollow region. The soil of the latter is principally black slate, with outcropping boulders of marble and limerock. In 1899, Mr. W. W. Eggleston of Rutland had visited this valley, and reported the rare Rue-Wall Spleenwort and the Purple-stemmed Cliff-Brakes growing abundantly on Gregor Rocks. I followed his path of cliff-climbing, as nearly as possible. Now that the orchid season was practically ended, I was giving my attention to hunting ferns, and I knew I should find them among the limerock cliffs. I had recently collected the Walking Fern in its native haunts. I proceeded up the valley of Wash-Tub Brook, passing the limestone mills northward, toward Mount Anthony and Peckam's Hollow. Another stream, known as Hemlock Brook, flows down from the eastern slopes of Perkins' Hill, and joins the Wash-Tub stream near the lime crushers. All the streams in this western corner of Pownal flow to the Hoosac River, while the streams from the northern summits of the Dome and Mount Œta flow northward to the Walloomsac.

For the greater part of the valley, Wash-Tub Brook flows through open pasture lands. The bed is broad and shallow, strewn with numerous small limerock boulders drifted down from the hills with the floods of spring. The larger boulders wear scars and dimpled erosions of the glacial period. I took time to explore the ledges above, where the depressions reveal the terraces of an ancient lake. The prevailing evergreen trees here appear to be hemlock and cedar—the American *Arbor Vitæ*—whose roots cling to the cliffs, their green spires lending a touch of coloring to the barefaced walls.

I saw from the banks here the distant pot-hole formations in the brook, from which the stream had taken its name. As I approached these marble basins, I found three in succession—one above the other—following the course of the stream—a narrow passage eroded through crystalline marble and limestone. The first or lowermost one was like a small lakelet overflowing its brim. The second one was a typical pot-hole, revolving its stones in its whirling waters. The bowl was about six feet deep, of a circular—or rather elliptical form, about twenty-six feet in circumference. The stream entered through the middle of the northern rim, and had eroded a spout-like gutter, causing the water to flow in a rapid, seething manner as if poured into the basin below. Here the greenish water boiled and whirled, finally with an added force leaping forth through a deep spout over the lower rim of the pot, carrying with it small stones and marble dust—the lower rim thus being worn away. As the bowl becomes deeper, layers of rock will be cracked and broken, until finally the pot-hole formation will be destroyed. The upper or third basin is located in the harder portion of the lime and marble bed-rock, portions of the marble being highly polished. The marble brook-bed glittered in the noonday sunshine. Pot-holes are formed originally by a boulder, which—carried in the currents of a stream—lodges in a depression of the bed-rock. It bores gradually into its resting place, until, in the course of ages, it has worn the walls of its basin into a deep hollow, at the same time wearing itself away, at last being carried off as a pebble. It may be that these holes are sometimes formed in a slightly different way. Dimples and fissures often occur in rocks, and if the water and pebbles circle about these cracks they probably eat down through the soft layers of rock, and thus loosen a revolving stone from the bed-rock itself. It would

The Gregor Rocks, Hoosac Valley, from Pownal Centre, Vermont.

then fit the pot-hole closely for ages, revolving as the currents become forceful in freshets. The potholes along the granite ridges in Bronx Park, New York City, as well as on the Canaan Hills—nearly one thousand feet above the Merrimac and Connecticut river-beds,—reveal the erstwhile revolving stones now motionless in their basins.

After remaining in the region of the Wash-Tubs an hour or more, I followed down the lateral moraine or wooded ridge along the stream, which became rocky in the heart of the hemlocks. Upon a broad table-like rock, I found a large mat of Walking Ferns. It appeared about four feet square, and contained the most luxuriant plants I have ever seen or expect to see. I placed several in my vasculum, and descended to the stream, hastening on toward the village. Here I met an old gray-haired man—the inn-keeper for the mill laborers. He recognized my botanizing outfit, and remarked that Mr. Eggleston had passed through the town in 1899. He directed me to the Gregor Rocks, above the village, and thus I found the path winding around the northern brow of these limerock cliffs.

Crossing the Pownal Centre road, I entered the pasture east of the village church, and wound up the cliffs above the limekiln quarry. Here, striking in among the cedar trees and ragged bluffs, I pulled myself up under the trees and rocks. Resting for a moment, I beheld a fern which proved to be the Purple-stemmed Cliff-Brake (*Pellæa atropurpurea*). Much elated by my discovery, I fell to wondering what the little Wall-Rue Spleenwort (*Asplenium Ruta-muraria*) could look like. I had studied the plates of these rare ferns, but had not known them face to face. I soon came to an enormous limerock boulder on the summit of Gregor Rocks, and here I found the rare ferns for which I searched. From the crevices both the tender green tufts of the Wall-Rue Spleenwort and the wiry purple stems of the Cliff-Brake grew luxuriantly, draping the fissured sides of the boulder. Climbing to the top of the boulder, I saw beyond my reach a tuft of Walking Fern. This proved to me that this plant throve on the dryest of limerocks in the fall glare of the sun, as well as in damp sheltered places. None of these species look like the ferns that are ordinarily known, and unless one turns the leaves over and observes the *sori* or fruit dots, he would never guess to what family they belonged, so different are they in appearance from their brothers of the boglands and hillside pastures.

The rocks about were covered with tufts of the delicate Wall-Rue, and great tangles of the Cliff-Brake, growing from twelve to fifteen inches tall. The last year's growth was still brown and rusty amid the fresh green fronds of this season. Hardy indeed were these ferns, growing in such a dry, exposed place.

Later in the month I made another trip to secure some ferns for photographing. It was Sunday, and the church bells were ringing at North Corners as I drove into the valley, and hitched my horse opposite the village inn. As I went my way toward the haunts of the ferns, I soon discovered that I was not making my ascent to the cliffs alone. A

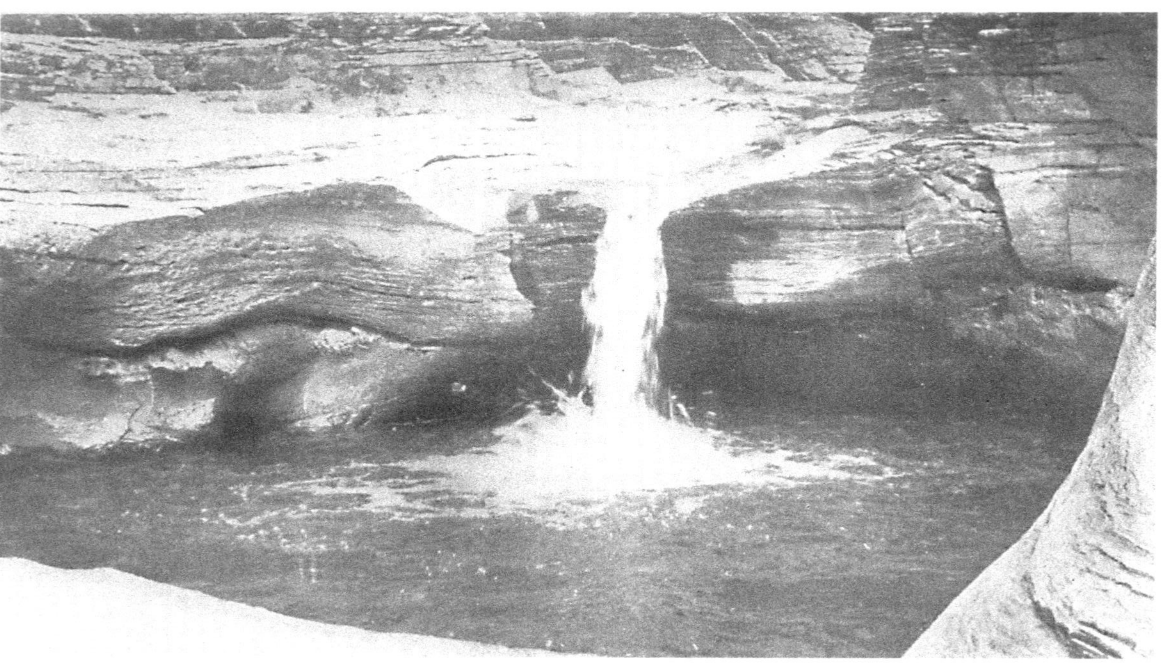

The Pot-Hole of Wash-Tub Brook, Pownal, Vermont, Showing the Stream Whirling through its Basin.

An Ancient Pot-Hole, Showing an Erstwhile Revolving Stone, Located on the Granite Ridge, near the Wolf's Bronx Park, New York City.

"The finest workers in stone are not copper or steel tools, but the gentle touches of air and water working at their leisure with a liberal allowance of time."—Thoreau.

"The stones which completed their revolutions perhaps before thoughts began to revolve in the brain of man. The periods of Hindoo and Chinese history, though they reach back to the time when the race of mortals is confounded with the race of gods, are as nothing compared to the periods which these stones have inscribed. That which commenced a rock when time was young, shall conclude a pebble in the unequal contest."—Thoreau.

gray-haired woman, with basket on her arm, overtook me. She seemed to be gathering the bluebells along the ledges. We began to converse, and when we came to some ripe strawberries, we ate in a social way the fruits we found by the path. She told me she was gathering bluebells to decorate the chancel of the church, as it was Children's Day.

On the brow of Gregor Rocks I asked my companion if the legend were true, of which Hawthorne writes, in 1838, during his stay in the valley at North Adams: "A mad girl leaped from the top of a tremendous precipice in Pownall, hundreds of feet high, and, if the tale be true, being buoyed up by her clothes, came safely to the bottom."[1] She told me the name of the girl who had made the leap. She was a half-witted creature who, descending the cliffs at twilight with a package of wool rolls, thought to save time by throwing her burden ahead of her and leaping from the rocks. Her homespun garments caught and held her in the cedars below, until the villagers heard her screams and rescued her. The rocks are called "Weeping Rocks" —for what reason it is not quite clear, unless through some exaggeration of this story.

I collected some perfect ferns, and told my companion their names. She glowed with interest, and told me she had never been to these cliffs since she was a child, until now. She said if she had her life to live again, she would have devoted more time to exploring these rugged hills and vales. Soon our baskets were filled, and with a warm handclasp we parted.

I proceeded up Wash-Tub Brook, and secured some fresh plants of the Walking Fern in Hemlock Glen; then I returned to my horse. I was laden with rare treasures from the roadsides before I reached Mount Œta, late in the afternoon.

Many of the present names of ferns, lichens, and mosses originated with the ancients. Dioscorides knew and designated two kinds of fern. They were thought to put forth no seed in those days, since they produced no flowers—except as Dodoens in 1578 wrote:

"We shall take for seede the blacke spots growing on the backside of the leaves, the which some do gather thinking to worke wonders, but to say the truth, it is nothing else but trumperie and superstition."[2]

The Osmunda, Polypody, Oak-Fern, Hart's-Tongue, Spleenwort, Asplenium, Venus-Hair and Maiden-Hair, as well as the delicate *Ruta-Muraria*—the Wall-Rue found on Gregor Rocks—were described clearly by the earliest herbalists. These records are full of errors and confusion, since the natural affinity of these species was not then known.

The Lichens were known also as "Stone-Liverworts (*Hepatica*), found with wrinkled, crimpled leaves on the ground or moist sweating rocks, where the sun shines seldom," according to Dodoens. Among the list of mosses described, I discover that our Round-Leaved Sundew, the little carnivorous plant, was anciently classed as a species of moss, in close relation with the Ground and Club mosses known as *Lycopodium*. But the Sundew, unlike the mosses, produces a stalk with

The Bluebells of New England. (*Campanulæ rotundifolia.*)

The roses are a regal troup,
And modest folks the daisies;
But, Bluebells of New England,
To you I give my praises.

Thomas Bailey Aldrich

white flowers. The plant was considered strange, because the stronger shone the sun upon the round, reddish leaves, the more moist with drops of dew became the plant; for this reason it was called in Latin, *Ros Solis*, which became in English *Sundew*, in 1578.

The Wall-Rue Fern was thought to resemble the Garden Rue, but is much smaller. Rue-of-the-Wall was common in Germany and England in 1578, and was found upon old moist cathedral walls where the sun did not shine. It was originally called, in apothecary shops, *Capillus-Veneris*, *Adiantum*; and in France *Saluia vita* and *Ruta-Muraria*. There were two varieties of this fern, designated in Europe as *Venus-Haire* or *Lumbardie Maiden-Haire*, in 1578. The larger species grew commonly about well-springs, in walls in Italy. It was known as *Capillus-Veneris*,—named by the ancients *Adiantum*. This fern has hairy foot-stalks, small, blackish leaves, snipped around. This species is, no doubt, our Venus-Hair Fern, known to-day as *Adiantum Capillus-Veneris*.

The Walking Fern was known to Linnaeus by the name of *Asplenium*, species of this genus being used against diseases of the spleen and liver. It was unknown to Dodoens in 1578. The Purple-stemmed Cliff-Brake was originally known as a species of *Pteris*, a name suggested because these ferns resemble the wings of birds.

Our native species of Bluebells of New England are emigrants from Europe, and are closely allied with the Bellflowers of Europe. These flowers were likened to cathedral-bells, with a small white clapper hung in the middle. These were, according to Lyte, found in Coventry and Canterbury, England, 1578, opening after "Sunne-rising," and closing toward "Sunneset." Theophrastus knew these flowers centuries before Christ, while Pliny designated them in Latin *Iosione*.

Our Bluebell (*Campanula*) derived its generic name from *campana*, the Italian for a bell. The species found on Gregor Rocks are known as *Campanula rotundifolia*, signifying round-leaved bells. The original Bluebells of Europe were known in 1578 as *Campanula cærulea*, from whence the common name originated. That of the "Bluebells of New England" originated with Thomas Bailey Aldrich, in his poem on these flowers.[3]

Between the 15th and 19th of July I made journeys over the nearer hills. I visited Oak Hill above White Oaks Valley, where I found the bluebells abundant along the roadside walls, even growing in the dooryard fences of the dwellers thereabout. I had never visited this hill before, and was charmed with its "glen-like seclusion." It is situated in the shadow of the Dome and Mount Hazen, surrounded on the northern and eastern sides, by wild, primeval forests. The broad, sloping meadows were among the first to be cleared in this region, and wear the scars that follow in the trail of the woodman. In the time of the Rebellion, many slaves sought the seclusion of this valley, and built their shanties snugly by the brooks. Until quite recently, the roads of the Hollow and the streets of Williamstown were frequented by one of these ancient slaves, known as "Old Abe-the-Bunter," who used to sell huckleberries and arbutus, and who sawed wood for the students at Williams. His real name was Abraham Parsons. The title "Bunter" was affixed on account of a horny growth projecting from his forehead, which he used sometimes after the manner of a goat. At one time, many years ago, a number of students and White-Oakers made a wager with Uncle Abe, which he won by butting through the heavy oak head of a molasses hogshead. It is also reported that after this, some students, putting a grindstone into a sack, told Uncle Abe it was a *tough cheese*. The old negro gave it a terrific bunt and cracked the stone, but nearly killed himself by the operation. He is said later to have killed a horse with which he had become enraged, by one blow of this horny growth. Carroll Perry has published a college book in which Uncle Abe figures in one chapter. It is entitled *Bill Pratt, the Saw-Horse Philosopher*.

Civilization and the selling of the streams for the North Adams water-supply has caused the removal of all the shanties along the Hollow Road. Only the old George Adams cottage remains as an example of the original type.

The region of White Oaks formerly included all the rocky hills and swamps now known as Colesville and Riverside, and has received its name from the abundance of white oak timber in this locality, utilized by the colored people in making baskets which they peddle in town. Many years ago, three very large white-oak trees stood east of the house known as Old Stone Tavern, near Broad Brook bridge. This building still stands—in a deplorable condition—as a tenement house. It is over a century old, built in the Revolutionary days by Silas Stone, who kept tavern when stages ran between Pittsfield, Bennington, and Troy.

On July 18th I made a tour of the limestone ridge above the Gulf Road, known as the Glebe. It is the most desolate and unproductive soil in the whole town. In the early townships of the New Hampshire Grants, Governor Benning Wentworth required "One Share for a Glebe for the Church of England as by Law Established," and another for schools. These lands, therefore, are to-day called "minister lots" and "school lots." The occupants, instead of paying taxes, pay lease money for the use of the land, which is appropriated according to the vote of the town's people for the support of ministers and schools.

Beginning northwest of the Swamp of Oracles, over the Amidon fields, one finds the limestone bed-rock cropping out everywhere. Little rounded hills appear to jut out of the deeper swamps leading toward Iris Swamp on Ball Farm, as one rides along the Pownal Centre Road. Great lime-rock boulders and piles of loosened rock lie strewn over the fields. One enormous boulder may be observed by the road north of Amidon's house, and another near the Peleg Card house. I collected innumerable Walking Ferns scattered over these miniature hills and boulders. I proceeded northward to the Campbell

Three Rare Ferns from Gregor Rocks and Wash-Tub Brook Region, Pownal, Vermont 1. Rue-in-the-Wall Spleenwort. (*Asplenium Ruta-muraria.*) 2. Purple-Stemmed Cliff-Brake. (*Pellæa atropurpurea.*) 3. Walking Fern. (*Camptosorus rhizophyllus.*)

horseshoeing shop, in the woods beyond, and turned to the left toward the ridges of the Glebe. Searching the rocks along the edges of the road, I found—perched high on a point of rocks—a beautiful colony of the rare Ebony Spleenwort (*Asplenium platyneuron*), not common hereabout. Over the mossy rocks below I again found numerous mats of Walking Fern. In finding these two ferns so closely associated, I searched for the rarer hybrid of these ferns, known as Scott's Spleenwort (*Asplenium ebenoides*), but did not find it. It has been seen but once or twice in Vermont, to my knowledge, being more frequent in southern New England, Alabama, and Virginia, where it ascends fourteen hundred feet above the sea level.

The rich soil amid the hollows above was covered with the strange Grape-Fern, locally called Umbrella-Fern (*Botrychium Virginianum*). Maiden-Hair and numerous other common ferns and brakes filled the swamps below. Coming from the woods, I entered a hayfield where the mowers were at work. Beyond this, I entered a cow-pasture skirting the Glebe ridge. Here were deep hollows guttered out, leading northward to Pownal Centre. Pennyroyal grew over the parched, dry plains, and in the hot sun shed forth its aromatic perfume. Boulders and natural obelisks were lodged on the hills above. In character the latter are similar to rocking stones, that are so finely poised on the mutton-backed bedrock, that with pressure they sway slightly. The obelisks are either pillar-like boulders moored in the mud and soil, or formed along cliffs by the heat, frost, and wind erosions, causing them to appear like columns or broken monuments, in the distance.

On the rocks of the Glebe hills, I again collected the Walking-Fern, and I am sure that if I were to penetrate the cliffs of the gulf along the western slopes of this ridge, I should find the Rue-of-the-Wall and Purple-stemmed Cliff-Brake.

Far away in the hollow, slept the little village of Pownal Centre. The church steeple towered among the trees, and the village green sloped between the church and the old Revolutionary road.

Notes:

[1] Hawthorne's *American Notes*.

[2] Dodoens, *History of Plants*, p. 290. 1578.

[3] T. B. Aldrich, *Bluebells of New England*.

From photograph by George Stonebridge.

The Rocking Boulder, Located on the Granite Ridge near the Bear's Den, in the Zoological Garden, Bronx Park, New York City. A pressure of fifty pounds causes this boulder to move about two inches.

"These, and such as these, must be our antiquities, for lack of human vestiges. ... The walls that fence our fields, as well as modern Rome, and not less the Parthenon itself, are all built of ruins."—Thoreau.

16
Alpine Blossoms of the Dome

> Mountains seem to have been built for the human race, as at once their schools and cathedrals; full of treasures of illuminated manuscript for the scholar, kindly in simple lessons for the worker, quiet in pale cloisters for the thinker, glorious in holiness for the worshipper.—John Ruskin.

Of the swamps and domes of the Hoosac region, Henry Ward Beecher once said: "The most level portion of this region, if removed to Illinois, would be an eminent hill. The region is a valley only because the mountains on the east and west are so much higher than the hills in the intermediate space. The endless variety of such a country never ceases to astonish and please. At every ten steps the aspect changes; every variation of atmosphere, and therefore every hour of the day, produces new effects. It is everlasting company to you. It is, indeed, just like some choice companion of rich heart and genial imagination, never twice alike, in mood, in conversation, in radiant sobriety, or half-bright sadness, bold, tender, deep, various."

On July 19th I drove beyond the Bogs of Etchowog, over a portion of the Hill Road toward Bennington. As I passed the Elijah Mason Farm, I turned my horse's head through the cow-pastures to the east. In a swamp to the left of the grassy wood-road, I collected scattered Pogonias and Limodorums, although the season was late for them. Still farther eastward are impenetrable swamps, through which Ball Brook flows northward to the Walloomsac near Bennington. The road led around to another swamp farther eastward, toward which I drove. It was one of those wild regions, tangled with tamarack, balsam-firs, high-huckleberry trees, amid the peat and sphagnum. The green spires of tamarack and fir swayed in billowy waves as the wind breathed through these vales; and the sunshine drew forth the fragrance of pitch and balsamic resins from the blistered bark of these young trees.

I fastened my horse to a pine tree, and penetrated the depths of this swamp as far as I dared, along a moss-grown brook bed leading from a spring toward the interior. The heart of this region was impenetrable. The pioneers, settling along the valley of Ball Brook, chose in Revolutionary days this heavily timbered region, in preference to the lower swamps of the deeper vales of the Hoosac. It has proven to be the coldest, most desolate, and barren soil for corn and grains,—the most productive crops here being stumps and boulders! Shad-bushes and the high-huckleberry bushes were laden with berries. I stood upon a log and ate of them for some time, meanwhile listening to the choruses of locusts and numerous thrushes, screaming jays, young crows, and whistling hawks. Many distant sounds came whispering to me from out this wild solitude of Nature. The mystery of wild wood isolation, in the presence of the scars of ages, took possession of me, and filled me with a nameless fear. I gave vent to a wild howl in order to relieve the tensity and portentousness of the situation. It was a damp, mossy place, such as bears, lynxes, and wild cats choose in which to nap during the day, being located in their run from the Petersburgh Hills to the Dome of the Green Mountains eastward, above. As Thoreau described one of the Maine woods swamps: "It was ready to echo the growl of a bear, the howl of a wolf, or the scream of a panther; but when you get fairly into the middle of one of these grim forests, you are surprised to find that the larger inhabitants are not at home commonly, but have left only a puny red squirrel to bark at you. Generally speaking, a howling wilderness does not howl: it is the imagination of the traveler that does the howling."[1]

I ventured on farther east, until I came to the true spring of the swamp. Every swampy region reveals innumerable springs, and this swamp was no exception. Many were oozing through the carpets of moss. Around such fountains I searched for the familiar leaves of Moccasin-Flowers without success.

I returned to the open pastures, all fear of the wilderness having subsided. I looked about, and saw from the lay of the land that this had been the bed of a glacial lake. It is in such regions as these that fossil remains of the whale and mastodon have been found. A fossil whale was found in Charlotte, Vermont, sixty feet above the level of the lake and one hundred and fifty feet above sea level. In Swanton, in a ledge of rock blasted through for railroad purposes, a large deposit of fossil marine shells was found. Also fossil bones of the elephant were found in Brattleboro.

Beyond the Howling Swamp, an interesting glacial hill rises, dividing the swamp from the broader valley of Ball Brook beyond. The lower southern brow of this hill had been eroded by the currents formerly flowing over the ridge when a larger lake existed here. From the summit of this hill, one becomes conscious that not so long ago wide waters spread about. Two currents are evident,—one from the glaciated Dome, flowing westward, and one from the ice-capped heights of Mount Anthony, southeastward; the two currents mingling and rushing westward over the Glebe toward Pownal Centre and the natural dam at Gregor Rocks, toward the Hudson Valley and the sea.

Slowly—as the dam in the valley broke away and let the ice-currents out—the mountain lakes were drained off, and left these bare, round hills and deep, swampy hollows, where as soon as the climates grew temperate, forests of evergreens sprang up and flowers bloomed. Northward, toward Bennington, as far as the eye can see, one discerns a chain of rounded wooded hills and intervening swamps.

On my way homeward, I stopped at the Swamp of Oracles, and decided to climb up the sides of the ravine for a look at the Large Round-Leaved Orchis, found here in June. I passed through Clintonia Hollow, beyond the woodchuck's home, where I had observed the Small Round-Leaved Orchis in the little animal's dooryard. There I struck out westward up the hillside. I frightened up the same mother whippoorwill that I had disturbed earlier in the season. The little birds of the second brood were now large, and commencing to feather. They were fluffy, and of a dead-leaf yellowish-brown color. Their large, round, brown eyes were like small shoe buttons. They began to run about at sight of me. The mother, meanwhile, feigned a broken wing and moaned piteously, with actual tears in her sad eyes. I lifted the downy balls in my hands. They snuggled without fear in my sleeve, and closed their sleepy eyes. Finally I put them on the leaves together, and promised the mother I would not again disturb her.

We have two species of the Goatsucker Family (*Caprimulgidæ*), including the Whippoorwill (*Antrostomus vociferus*), and the Southern Whippoorwill, or Chuck-will's-widow (*Antrostomus Carolinensis*). The closely allied Night-Hawk, or Bull-Bat (*Chordeiles Virginianus*), is often mistaken for the Northern Whippoorwill. Its habits and flight are far different, however, although the homes of both are similarly adopted. The Night-Hawk deposits her two buff-green eggs on rocks, bare ground, or on flat roofs, either in country or village. All of these birds winter in the Southern lands, and all save the Chuck-will's-widow arrive here about the third week in May, returning with their broods the latter part of September.

The Twilight-Hawk preys upon other birds and moths. I have observed him at twilight, on a cloudy day in autumn, circling and diving down among the weeds about a potato field, where sparrows were feeding in great numbers. The sparrows flew in fear toward the house, one driving so forcefully against the window-pane that he dropped to the ground with a broken neck. This Night-Hawk gives forth a peculiar moan or call,— "Peent,"—accompanied by a booming, buzzing sound in flight, as the wind passes through the quills of its feathers. It whizzes swiftly through the air, swooping down upon its prey about the fields or garden.

The leaves of the Pink Moccasins—sometimes called Whippoorwill's-Shoes—were numerous about the place, the flowers serving, near the ever-changing nests, to attract the insects and moths upon which the birds feed.

I found another oven-bird's thatched nest in Witch Hollow region, late in June, very near the colony of Ram's-Head Cypripediums. On my return to secure a photograph of it, I found that some animal—perhaps a dog or skunk—had torn the nest to pieces and devoured the birdlings.

The Small Round-Leaved Orchis, which formerly I observed in Chalk Pond region, has developed into the varietal form of this species—producing oblong leaves—known as *Habenaria oblongifolia*. This often occurs when the flower is in company with the true Round-Leaved Orchis. This season I have instanced the fact in another colony of these orchids, in Rattlesnake Swamp.

The flowers found on the summit of the Dome, three thousand feet above sea level, are slightly modified in size and coloring. They are fully two or three weeks later in blooming than the same species flourishing in the Hoosac Lowlands.

On July 20th, with two other mountain climbers, I started from the brow of Mount Œta, and at nine o'clock descended to Rattlesnake Swamp and the secret haunts of Showy Reginæ. We crossed the stream over the log bridge, and followed up the old Joe Larabee path; passing around the southern ledge of the Domelet to the Dummy Road watering-trough. The path was densely overgrown with bushes, and impeded with heaps of treetops. However, we finally came out to the Exford Clearing and the White Oaks Road beyond. At the watering-trough, a road turns to the right hand through Rocky Hollow, leading to the Coal-Bed, or Chip-Bed, as it is known. We sauntered along the shady path of the Hollow until we came to the clearing, where loggers in winter haul and pile their spruce and hemlock logs for later milling. From this station, four roads branch in various directions. We took the northeast path, and were soon climbing steadily toward the clouds. On a previous occasion, during March, I had ridden on a logger's sleigh to the summit. The snow, then about four feet deep, covered fallen trees, over which, during summer, it is almost impossible to walk. In winter, the hardened, encrusted snow spreads a clear, smooth surface for walking, far above impassable barriers and tangled brush. In its summer garb, the road was strangely confusing to me. It was rocky, and intersected by sun-dried brook beds, which the melting snows had guttered in spring.

Rocky Hollow Road is available for horse and carriage as far as Logger's Depot, and northward to the Dummy Road. The trees along this vale are chestnuts, beech, yellow, white, and black birch, white oak, black oak, maples, and various flowering bushes, such as azalea, mountain laurel, and shad trees. As one ascends, the trees become dwarfed and gnarled, and many abnormal forms occur among the yellow birch. As we neared the summit, the yellow birch trunks assumed great size, while their tops were scraggy and dwarfed by the winds and storms. Higher up, we found little but spruce, hemlock, and balsam-fir; the trees and bushes became low-lying,—hugging the rocks for protection from the winds.

The Red Wood Lily. (*Lilium Philadelphicum.*)

O lilies, upturned lilies,
How swift their prisoned rays
To smite with fire from Heaven
The fainting August days!

Elaine Goodale.

Frequently we paused by the path for breath, finding sweet Canada Violets (*Viola Canadensis*) ripening their seed-capsules. They were ready to burst and throw their seeds about for some feet. We collected several plants to transplant. The brakes and sphagnum indicated a swamp not far distant. We began to feel thirsty, and searched about without finding trace of a spring, although one is said to be near here, with a rusty tin cup hung to a tree. To the left of the path, we saw the ruins of a woodchopper's log cabin, which assured us that brook or spring must be near, else the spot never would have been chosen for man's habitation. Above the hut, we came to a clearing. A level stretch led to the junction of two roads: one led directly ahead, terminating on the Ladd Lot, while the path to the right turned abruptly up the steeps to the summit of the Dome. The last few rods were the steepest portion of the whole journey, the rest of the climb having wound around about in gradual ascent.

At last we walked along the edges of a precipice above Bear Swamp. In the scorching heat of noon we made one last turn eastward, entering the clearing on the very brow of the desolate Dome, three thousand feet above the sea. Here were dense groups of beautiful spruces and balsam-firs. The forest floor was carpeted with luxuriant leaves of clintonia and dwarf dogwood,—sometimes wrongly called bear-berries. The latter, an Alpine species, was still in bloom, the flower sometimes having two whorls of rosy-tinted petals. The mountain snowberry, creeping wintergreen, trailing arbutus, and goldthread were clinging to the sphagnous hummocks over the summit, while Alpine species of huckleberries crept through the clearing and draped the white-faced rocks.

The great stillness of Nature's solitude was broken only by the buzzing of insects, the notes of the chickadees, and the winds soughing through the boughs of spruce and firs. The brow of the Majestic Dome receives the force of the eight winds of heaven direct from the frozen North or from the fragrant Southlands. In March, 1894, a terrific tornado swept over this region from the northeast, mowing a path several rods wide over the Dome, and laying the spruce and firs in a twisted pile;—that portion of the summit is almost impassable today. During these great northeasters in the spring, the birds and beasts of the Dome seek the lower plains and hollows.

We wandered southward in the path of the tornado, a quarter mile or so, to a sphagnous swamp and the ledge of White Rock on the side of the Dome. The view from these rocks is variable, yet not picturesque nor pastoral as the one from Mount Œta. It is wild, fearful,—beyond all signs or sounds of civilization. Far to the southwest the blue Catskills blend with the sky; southward the grim, awkward, ragged shoulders of Greylock's Brotherhood tower; from the eastern brow, Haystack and Stamford Mountains roll away, one after the other, like great land waves. The deep valley of Broad Brook sleeps below. The slopes of Stamford Mountains are dotted with evergreen trees for miles, as far as one can see.

Gathering a few fragrant balsam-fir boughs, we now rapidly began to descend the mountain, for while the luncheon we carried had satisfied our hunger, we were sadly in need of drinking water. We soon found ourselves at the Coal-Bed, gathering the Wild-wood Tiger Lilies (*Lilium Philadelphicum*), which we had observed as we passed in the morning. We ate the late wild strawberries along the roadside, and took a long rest in the shade, pursuing our way later down the Rocky Hollow Road northward to Blackberry Clearing, on the Dummy Farm. Here we religiously searched the ravines for Deaf-Man's Spring. Major, our dog, was the first to discover it. We found him taking a bath in the deepest pool. However, a higher basin was overflowing with fresh, clean water, from which we drank excessively. The reviving effect upon our spirits was immediate. Deaf-Man's Fountain is in the ravine of Dry Brook, walled up like a little well. It is the only water in this immediate vale,—a natural and everlasting spring-head. Guide-boards should be erected at the four corners of country roads, directing travellers to the water-supply, the need of which is often so powerfully felt by pilgrims.

We rounded the Domelet, descended to Jepson Farm in Rattlesnake Valley, and proceeded to Lloyd Spring and the colony of Showy Reginae. At this point in our travels, we had completed a great circle.

Notes:

[1] Thoreau, *Maine Woods*, p. 300.

17

The Cascade and Bellows-Pipe, Notch Valley, Berkshire County

Come here where Greylock rolls
Itself toward heaven; in these deep silences
World-worn and fretted souls
Bathe and be clean! Cares drift like mists away.
—Author Unknown.

Monday, July 22d, dawned fair, although there were some signs of a storm in the lowering gray cloud-folds at the horizon. However, we had decided to explore the Notch Valley and the Bellows-Pipe, between Greylock and the Ragged Mountains.

We journeyed from Mount Œta to North Adams, leaving State Street about ten o'clock, and ascended the path to Witt's Ledge. Soon we rounded the Ragged Mountains, entering the woods near Crystal Spring, where we descended the Cascade ravine. Its rocky chasm is beautifully draped with the Common Polypody Ferns, and delicate tufts of Maiden-Hair Spleenwort, which clings in the fissured ledges. The bed-rock appears to be a flinty slate, similar to that of the Tunnel Mountain. It is not so favorable to the growth of the rarer ferns—such as the Rue-in-the-Wall—as the lime-rock formation of Gregor Rocks in Pownal. Large boulders lie in the heart of the brook bed, and the hillsides are clothed with primeval hemlocks. Just above the brow of the Cascade, I found a few Walking Ferns. The ravine is accessible to this point, but here I was forced back and climbed the southern bank to the path leading around to the waterfalls. High boots supplied with hob-nails are indispensable to safety in such climbing in the channels of streams.

From this point we retraced our steps to the Pent Road, leading up through Snuff Hollow to the City's reservoir, at the junction of the South Adams Road. Here we trudged up the hill and entered the Notch highway at Walden's farmhouse. Greylock Park Road turns off here through the pastures, around Mount Williams. We, however, continued straight ahead toward the source of The Notch Brook,—Hawthorne's and Thoreau's routes, long before roads to Greylock were available. It was steady climbing, until at last we reached the pasture-

The Cascade of Notch Brook, at the Base of Mount Greylock's Brotherhood, North Adams, Massachusetts.

The highest lands of Berkshire's noble hills
Shall sweetly ring with song and louder trills;
And many a spring within the Bellows dumb
Shall swell and flow with swift, yet soothing hum.

G. G. N.

land where the streams from Greylock's Brotherhood divide; there is a stream beyond the ridge, flowing southward to South Adams, while those on the north side flow down Notch Valley to the Hoosac River. Hawthorne often sought the seclusion of this valley, and in his *American Notes*, under date of September 9, 1838, describes these rugged slopes. He not only followed up the North Notch, but descended the South Notch in the rocky course of the stream homeward through South Adams. He speaks of inquiring at a cottage his way to South Village, which was "across lots," into the road near the Quaker Meeting-house, surrounded by grave-stones. He also drank of the region's spring water,—the "most delicious" he ever tasted,— "pure, fresh, almost sparkling, exhilarating,—such water as Adam and Eve drank."[1]

The people of this region looked upon his journeys through their valley with curiosity in those early days. The houses were more numerous then than now in the extreme southern portion of the valley. This region has been purchased by the North Adams Water Company, which has removed all dwellings above the reservoir. The last house in The Notch to-day is on the Walden Farm, at Greylock Park Gate.

Hawthorne found, in the Highlands-of-the-Hoosac, the originals of many characters described in his works. "Eustace Bright," of *Wonder-Book*, was a student of Williams; and the *Tanglewood Tales* have made the whole world familiar with "rough, broken, rugged, headlong Berkshire." Here, in the seclusion of The Bellows-Pipe, "where it slopes upward to the skies," Hawthorne loved best to come. There he could look southward over the vast fields of Berkshire's valleys to the distant crags of Bryant's "Monument Mountain," immortalized as the "headless sphinx" of his own *Wonder-Book*. And from the northern Notch, he looked away to the blue Domes of the White Mountains, a distance of sixty miles or more.

The Limekilns along the Ashuilticook—the south branch of the Hoosac—still are smoking, as when Hawthorne and Mr. Leach visited them in 1838. The tale of *Ethan Brand* was suggested by the legend of an insane creature who threw himself in at the open gate of the burning kiln. Their open iron doors in the mountain-side at night seem like yawning mouths of Tartarus. Hawthorne met here also his "Bertram," who figures in the story; while "the boy Joe," son of "Bertram, the lime-burner," was a bar-room lad observed at the "Whig Tavern" in North Adams. Daniel Haines, then living in a desolate hut in "Willow Dell," was formerly nicknamed in the village as "Black Hawk," and is described in *Ethan Brand* as "Lawyer Giles," the "elderly ragmuffin," who—with the rest of the lazy regiment from the town tavern—came in response to the summons of "boy Joe" to see poor Brand returned from his long "search after the Unpardonable Sin." The title of this story was the name of one of the prose master's Salem acquaintances.

Among other characters which Hawthorne drew from this region, were the "seven doctors of the place." In the "Whig Tavern boarder"

Hawthorne saw and delineated himself. He describes the Saddleback Mountain and Greylock in all their different phases,—when enshrouded with dark masses of storm clouds and when: "Old Greylock was glorified with a golden cloud upon his head. Scattered likewise over the breasts of the surrounding mountains, there were heaps of hoary mist, in fantastic shapes, some of them far down into the valley, others high up toward the summits, and still others, of the same family of mist or cloud, hovering in the gold radiance of the upper atmosphere. Stepping from one to another of the clouds that rested on the hills, and thence to the loftier brotherhood that sailed in air, it seemed almost as if a mortal man might thus ascend into the heavenly regions. Earth was so mingled with sky, that it was a day dream to look at it. To supply that charm of the familiar and homely, which Nature so readily adopts into a scene like this, the stage-coach was rattling down the mountain-road, and the driver sounded his horn, while Echo caught up the notes, and intertwined them into a rich and varied and elaborate harmony, of which the original performer could lay claim to little share. The great hills played a concert among themselves, each contributing a strain of airy sweetness."[2]

As we neared the head of The Bellows-Pipe, and passed the Wilbur and Eddy farms, where Thoreau was entertained, I tried to trace the paths which he had followed in his ascent to Greylock some years after Hawthorne sojourned here. He stopped that July afternoon in North Adams Village, purchased a tin cup, a little rice and sugar, and, placing them in his knapsack, started up The Bellows toward the mountains, followed closely by a thunderstorm. "The thunder had rumbled at my heels all the way," he said, "but the shower passed off in another direction, though if it had not, I half believed that I should get above it." He "reached the last house but one, where the path to the summit diverged to the right, while the summit itself rose directly in front." But it seems he "determined to follow up the valley to its head," and there find his "own route up the steep as the shorter and more adventurous way." He believed this "occupied much less time than it would have taken to follow the path—for what 's the hurry? If a person lost would conclude that after all he is not lost, ... but the places that have known him, *they* are lost,—how much anxiety and danger would vanish. I am not alone if I stand by myself."[3]

We followed up the eastern sides of Notch Valley to the head of The Bellows where the Saw Mill had stood in Thoreau's day. We regaled ourselves upon the red raspberries along the pasture, and found the Deadly Nightshade in bloom amid the bushes. These fields furnish pasturage for yearlings and calves. The sides of Greylock are clothed with a heavy forest— "all beshaggled,"—and adorned with "headlong precipices" and innumerable rivulets. Finally we crossed to the west side of the valley, in the shadow of the great hill, and entered a ravine which we christened Æolian Glen.

I have always believed that this Notch Valley was in Thoreau's thoughts when he wrote "Rumors from an Æolian Harp." The name "Bellows-Pipe" originated with the early settlers for the extreme portion of Notch Valley, on account of the subtle roaring of the southeast winds, breathing like a bellows through the narrow vale. The Indians recognized in the roar of winds the anger of the Great Spirit. The Hoosac Highlands near the "Forbidden Mountain" were their hunting grounds, to which they journeyed from their Indian village farther westward near Schaghticoke, not far from Troy-on-the-Hudson.

Thoreau says of this vale's "glen-like seclusion overlooking the country at a great elevation between these two mountain walls," that it reminded him of the homesteads of the Huguenots, on the interior hills of Staten Island.

As Thoreau passed the last house in The Bellows, on his ascent to Greylock, "Rice" called out and told him that it was still four or five miles to the summit by the path which he had left, though not more than two in a straight line from where he was, but that nobody ever went this way; there was no path and it would be found as "steep as the roof of a house." But Thoreau took the short cut, notwithstanding Wilbur's warning that he would not reach the summit of Greylock that night. Thoreau says, however: "I made my way steadily upward in a straight line, through a dense undergrowth of mountain laurel, until the trees began to have a scraggy and infernal look, as if contending with frost goblins, and at length I reached the summit, just as the sun was setting." After taking "one fair view of the country before the sun went down," Thoreau "set out directly to find water." It proved to be labor, too. Following down the path for half a mile he came to a muddy place in the road "where the water stood in the tracks of the horses which had carried travellers up." He drank these dry, one after the other, by lying flat on the earth. He was not able to fill his dipper, and in a place above dug a well about two feet deep, using his hands and sharp stones as spade and hoe. It soon filled with pure cold water, from which he filled his tin cup; and he says: "The birds, too, came and drank at it." He then proceeded to the rude wooden observatory originally erected by Williams College, for the construction of which Platt— "a friend of mine," writes Hawthorne in the *Diary*—hauled the material by ox-team. Platt, the stage-driver, boasted of the fact that he was the *first* man to drive a team to the summit of the then pathless Greylock, led by President Griffin of Williams on horseback, who directed the building of that first observatory. This tower is now replaced by a modern iron structure fifty feet high.

Thoreau collected some "dry sticks, and made a fire on some flat stones" placed on the floor of the observatory for the purpose, and cooked the rice which he had bought in the village, eating it with a wooden spoon whittled out for the occasion. He was up at daybreak the next morning, and he has left a glorious description of sunrise on Greylock, as seen from the tower in the mists.[4]

Notch Valley and the Bellows-Pipe, North Adams, Massachusetts.
Mount Greylock towers up on the right, and the Ragged Mountains on the left hand.

There is a vale which none hath seen,
Where foot of man has never been,
Such as here lives with toil and strife,
An anxious and a sinful life.

Thoreau

The nights are very chill on these summits, even in July. There are now several log-cabins erected on Greylock for travellers to occupy, with stables for horses and keepers in attendance. The Catskills can be seen to the southwestward from this height.

Thoreau set his compass for a lake in the valley to the southwest, and descended the mountain by his own route, on the opposite side to that of his ascent.

My companions and I had climbed the slippery glen to where Thoreau commenced his ascent, and a tiny rivulet slipped over the rocks, which had formerly been dimpled with miniature pot-holes. Along the moss-grown banks, above the brook-bed, grew the familiar leaves of the Wild Ginger, while at the very entrance I discovered the Wild Black Currants (*Ribes floridum*), similar in taste and appearance to the cultivated species. The fruit was covered with bristles, and produced a disagreeable odor like that of the Wild Red Currants on the Dome—reminding one of a skunk.

At the entrance of Æolian Glen, a long log-like slab of rock lay upon the ground, strangely suggesting a petrified tree. Slowly we descended the western side of the vale, counting no less than twenty-two flowing brooklets, and four sun-dried brook-beds between Æolian Brook, at the head of the Bellows, and Walden Farm below. As we approached the meadows where the Wilbur Farm buildings formerly stood, we found a half-dozen spikes of the Ragged Orchis (*Habenaria lacera*) amid the damp grasses. This species I collected also later in the pastures of Rattlesnake Swamp, and found the pure White-Fringed Orchis along the roadside of Ladd Brook Valley in Pownal.

We now arrived at Crystal Spring, where we freshened up before entering the City in the "hollow vale" three miles below.

The formation of the Notch Valley was brought about by one of the successive terminal moraines flowing from the glaciated slopes of the ice-mountains farther northwestward, in the Adirondack region; while later the glaciated shoulders of Greylock's Brotherhood slowly melted, eroding the slopes with small ravines in which the numerous rivulets flow today. The continental ice rivers from the higher glaciers northward apparently culminated in tremendous and successive cascades above Notch Valley, eroding the deep-cut gorges between Greylock and Ragged Mountains. The general directions of these currents, below these waterfalls, were various, finally leading down to the ancient Hoosac Lake, and flowing with it through the natural dam, northwestwardly, to the Hudson Valley, and thence to the sea. According to Professor T. Nelson Dale, an ancient lake six hundred feet deep existed in the Hoosac Valley ten thousand years ago. Perhaps ten times ten thousand years ago, a greater glacial sea overflowed the Hoosac Tunnel Mountains, leaving the bald summit of Greylock alone towering above the waves. As the terminal moraines of the great ice-sheet slowly receded, the various cascades formed pot-hole erosions, in their descent on the Canaan Hills, above the Connecticut

Valley. Deerfield Arch was similarly formed by the force and chemical action of the eroding ice rivers, which flowed from glaciers, and wore through the wall of rock spanning the Deerfield Valley. Hawthorne compared this arch to "the arched entrance of an ancient church, which it might be taken to be, though considerably dilapidated and weather-worn. ... It was really like the archway of an enchanted palace, all of which has vanished except the entrance—now opens only into nothingness and empty space. ... This curiosity occurs in a wild part of the river's course, and in a solitude of mountains."[5] Dr. Wolfe says: "The summit of the arch and the water-worn pillars upon either side display 'pot-holes' and other evidences of erosion, and in the bed of the current lie fragments of similar attrite rocks which seem to indicate that at some period a series of arches spanned the entire space from mountain to mountain."[6]

Other erosions known as the "Twin Cascades" are found on the eastern slopes of Hoosac Mountain, above the eastern portal of the Tunnel, formed ages before the Hoosac Lake rippled in the "hollow vale" at North Adams. The Natural Bridge of the Mayunsook Valley is one of the greatest natural formations in Berkshire Highlands, and was also caused by erosions of the ice-currents ages ago.

On August 16th, this season, a great landslide occurred on the southern brow of Greylock, caused by a cloudburst. It began within a few feet of the summit, widening as the loosened soil slipped off the bedrock of the mountain. It swept down with velocity, becoming several rods wide as it reached the valley. It covered Gould Farm with earth, rocks, and logs gathered in its descent to South Adams, and the machinery in the mills in the village, three miles away, was crippled by the sand and water pouring in about the engines; the streets became canals, and boats were necessary to move about in. However, no lives were lost. The formation rock, from the base of Greylock, is laid bare in the path of this landslide. Six to ten terraced ridges, like stone stairs, are revealed in the ascent for some distance, indicating many ages in geological history. Here is evidence of those slowly receding seas and lakes as they drained from the summits down, stair by stair to the winding Ashuilticook River of to-day.

Notes:

[1] Hawthorne, *American Notes*, September 9, 1838.

[2] Hawthorne, *Ethan Brand.*

[3] Thoreau, Tuesday, *Week on the Concord and Merrimack Rivers.*

[4] Thoreau, Tuesday, *Week on the Concord and Merrimack Rivers.*

[5] Hawthorne, *American Notes*, August 31, 1838.

[6] Dr. T. F. Wolfe, *Literary Shrines*, 173, 1895.

18
The Natural Bridge of Mayunsook Valley, Northern Berkshire

> There's no music like a little river's. It plays the same tune (and that's the favorite) over and over again, and yet does not weary of it like men fiddlers.—Robert Louis Stevenson, *Prince Otto.*

A narrow vale winds away northeastward from the city of North Adams to Stamford, Vermont. A short walk from the terminus of the car line in The Beaver leads to the junction of Hudson Brook with the Mayunsook River. The Mayunsook is often called the Little Deerfield. It is the North Branch of the Hoosac, rising near Stamford Ponds, and draining the southern and western slopes above Stamford Hollow. The Greater Deerfield River rises also near these lakes, and drains the same mountains from their northern and eastern slopes, flowing around through Readsboro to Zoar, where travellers meet it as they pass out of the eastern portal of the Hoosac Tunnel into the Deerfield Valley. Thus, from their mysterious sources our turbulent rivers and mountain streams bring restful, cooling news from out the higher lands, where scarce the foot of man has been.

On August 7th I explored about Natural Bridge on Hudson Brook. I wore hob-nailed boots, and made a long day's excursion. Hawthorne knew and loved this wonderful natural feature of northern Berkshire, and here gathered many fancies, which he has woven into his tales. The chasm of Hudson Brook is described as the "Cave" in his *Notes.* His description of the ravine is the finest ever written.

Hudson Brook, tradition tells us, took its name from the hunter Hudson, who, one twilight, dragging homeward a deer he had killed, lost it in this chasm. He narrowly escaped following it himself.

The region is entered either by walking up the bed of the stream itself, or following around the road above Marble Quarry, just east of the chasm. The former is the more direct, but the latter a longer and safer way. In this instance, I followed the travelled highway. I proceeded up the stream where the erosions begin, and readily descended the ravine, following its course downward until I came to a beautiful marble basin or pot-hole formation, which very few see, since it is hidden under the wooden foot-bridge above the natural bridge of rock. Logs and immense rocks barred my way, and I was forced through dark fissures in my ascent to the sunlight.

The pot-hole was evidently the same pool of which Hawthorne wrote: "As the deepest pool occurs in the most uneven part of the chasm, where the hollows in the sides of the crag are deepest, so that each hollow is almost a cave by itself, I determined to wade through it ... there was an accumulation of soft stuff on the bottom, so that the water did not look more than knee-deep; but, finding that my feet sunk in it, I took off my trousers and waded through."[1] He visited this stream often: "The cave makes a fresh impression upon me every time I visit it,—so deep, so irregular, so gloomy, so stern,—part of its walls the pure white of marble,—others covered with a gray decomposition and with spots of moss, and with brake growing where there is a handful of earth."[2]

Hawthorne believed firmly that "a complete arch of marble, forming a natural bridge over the top of the cave," must have covered the whole chasm of the stream at an unknown period. The pot-hole, I am most certain, has been forded by few lads, and it is hardly probable that any other poet or prose master ever disrobed and bathed in its waters as Hawthorne did in 1838. The basin is from six to eight feet deep, with a beautifully rounded, highly polished brim. I christened this bowl "Hawthorne's Bath-Tub," and, unable to wade it, climbed out of the "Cave" to the light above. I, however, descended again to see the northern portal of the arch below the Bath-Tub. I was interested in the names painted high and low upon the marble rocks. Some visitors had evidently tried to place their initials as high as possible, while others more modest sought to write theirs as low, and in more obscure places. I regretted that I had not brought a pot of red paint and a brush to daub my own title there, with the ambitious crowd.

The stream, as it approaches the arch of the Bridge, is deep and of a dark green color. The chasm, from the top of the ledge, is about sixty feet deep, and the ravine three hundred feet in length. Geologists say that the ravine was formerly spanned by two ledges of rock, one of which is now in ruins. The piles of rocks in the chasm south of the southern portal of the arch are dazzling white, seen in the noonday sunshine. The fall of water, in its descent through the ravine, is about forty feet to the three hundred feet, so that the eddies play and whirl rapidly through the irregular bed. A wooden tile, or raceway, was hung high over the chasm, across a leaning crag of the original ledge,—conveying water power thereby to mills below. This old structure leaked, and as I descended the banks below, I saw some of the most gorgeous miniature rainbows spanning the depths, as the sunshine fell upon the mist near the arch.

A lad once made a wager with his comrades that he could cross over the ravine upon this wooden tile. The old weather-worn log was slippery with mould and mosses. In making his daring and perilous trip, the youth lost his footing, and fell headlong into the heart of the chasm. Following the fall, a terrific thunderstorm passed through the Hoosac, and night closed over the chasm. The next morning the boy's lifeless body was recovered. The wooden structure is now replaced with an iron tiling.

I passed on down the path on the west bank, until I reached an immense marble boulder, which was draped with dainty ferns and mosses. Little rivulets flowed from its sides, and climbing around to its southern brow, I was delighted to find many luxuriant plants of Walking Ferns—this making the fifth excursion in succession in which I had found this rare plant.

I entered the ravine below the boulder, and picked my way up the chasm to the southern portal of the arch, where I became wet through from the mist above, as I ventured to look through the cave. Returning, I found a path up the east bank leading to Marble Quarry and the mill below, where gravestones, door-stones, and various ornaments are manufactured. The most useful piece of work ever turned out here was, in my mind, the Williams College sun-dial tablet, which Hawthorne observed in 1838 as being as large as the top of a hogshead.[3] I have later discovered that this dial was placed near that old Astronomical Observatory on Consumption Hill, near the present College Library,—the first building of its kind erected in the United States, for the study of the worlds above, by Professor Albert Hopkins, in 1838. The bronze sundial was supported upon the marble table which Hawthorne saw at the quarry. Around it was carved in the soft marble the now dim inscription:

"How Is It That Ye Do Not Discern This Time."

This dial is now among the relics in the College Museum.

The overhanging crag, near the southern side of the arch, will in another half-century or more tumble also into the ravine. One large pine tree and many bushes, growing on this leaning tower, are plying their roots deeply in the marble fissures, and are slowly splitting the rock asunder. I have designated this pile Captain Skipper's Monument, in memory of him who recorded the last evidences of the Beaver Dam across this stream. Tradition says that the beavers labored centuries before the white man arrived in the Mayunsook Valley, building better than they realized, since they erected a dam which stopped the rippling flow of Hudson Brook. Originally, this stream flowed nearer the surface of the Natural Bridge. It is believed by some that the dam dogged the driftwood from the domes, and thus set the waters back. The force of the eddies, combined with the chemical action of the waters whirling among the logs and rocks, eroded dimples in the soft marble, until they wore the present archway through.

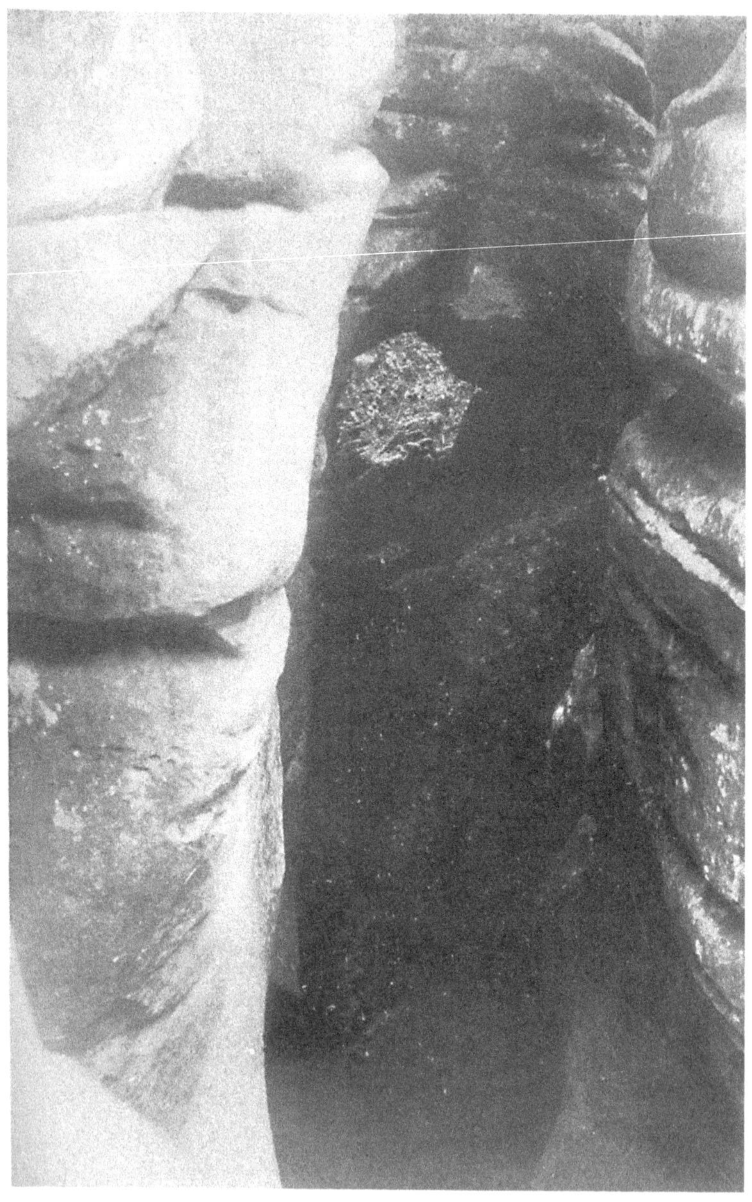

The Marble Arch of the Natural Bridge, North Adams, Massachusetts.

I followed down the bed of the stream, stepping from rock to rock easily, until I reached the path far below the Marble Quarry, and entered The Beaver, a little village where every one works like the small animals for which it is named. I was now near the junction of Hudson Brook and the Mayunsook; and not wishing to return to the City until sunset, I scrambled up the slippery sides of the hemlock hills above the little river. With the echo of the cavern's tumultuous roar still in my ears, I now heard, in pleasant contrast, the distant gentle murmur of that flowing stream. When I departed from the vales of these talking streams, I carried with me back to the busy world the remembrance of the voicing fantasies of their songs of wilderness and solitude.

Notes:

[1] Hawthorne, *American Notes*, July 31, 1838.
[2] *Ibid.*
[3] Hawthorne, *American Notes*, July 31, 1838.

"Our own country furnishes antiquities as ancient and durable, and as useful as any; rocks at least as well covered with lichens, and a soil which, if it is virgin, is but virgin mould, the very dust of nature. What if we cannot read Rome, or Greece, Etruria, or Carthage, or Egypt, or Babylon, on these; are our cliffs bare?"—Thoreau.

19
Orange Mountains, and Salt Meadows, New Jersey

> The weapons with which we have gained our most important victories, which should be handed down as heirlooms from father to son, are not the sword and the lance, but the bushwhack, the turf-cutter, the spade, and the bog-hoe, rusted with the blood of many a meadow, and begrimed with the dust of many a hard fought field.—Thoreau, *Excursions.*

August 12th I started for the Orange Mountains, in search of Cardinal Flowers, and various other blossoms, which I hoped to find about Eagle Rock. Arriving at those ragged cliffs, overhanging the brow of the mountains above West Orange, I climbed up the winding stone stairs and entered the park. The woods were strewn with small yellow flowers and ferns.

The view from the Rock is vast, as the eye sweeps off over the Great Salt Meadows beyond Newark, to Brooklyn Heights. On a clear day, the tall buildings of New York and the piers of Brooklyn Bridge are discernible. The Goddess of Liberty in the Bay also stands out clearly, and the slow-moving sails and funnels of outgoing steamers are visible. Most people seek Eagle Rock for this view alone.

Farther back in the woods, in May and June, the Pinxter-Flower, False Solomon's Seal, yellow and blue violets, bluets, and anemone everywhere decorate the rocky soil. Numerous tall weeds towered coarsely along the mountain-sides, to-day flaunting their disagreeable perfume ever before me.

I followed southwesterly, along the summit for a mile or more, to Crystal Lake, passing the park called "Wildmont," to the right of which stands Cobblestone Cottage. The building appears very ancient.

All the vast solitudes of the parks of Orange Mountains are locked within gates, and the entrance labelled, "*No Trespassing, Under Penalty of the Law.*" Law is a specific designation for a certain kind of a broad-headed, bow-legged quadruped—a thoroughbred species not mentioned in the scientific annals of the Hoosac Highlands. After passing the lake, I followed up the swamp toward the distant walls of

Wildmont, very desirous of trespassing and seeing the Wild Law in his cage. Soon I found a place where the stones were tumbled out, and where, by lifting a barbed wire, I could crawl through. So happily and leisurely I began to trespass about the woods. I found luxuriant colonies of the Maiden-Hair Fern, tall spirit-like spikes of feathery flowers, and club-like spikes of fringed-purple weeds not seen in the Hoosac Valley. They were so common that I did not gather any, so I never determined their title. In the deeper pools grew a few plants of the Skunk Cabbage. The low bushes and plants were overgrown and coarse in the extreme, amid the dense shades of chestnut and elm trees. The forest, apparently, was still in its primeval state.

As I approached the cottage of Wildmont, I ran upon an old cellar hole, where a building had once stood. The ruins were now prettily covered with myrtle and ivy. From this site, between the parting boughs, I caught glints of a sea of blues in the valley of the Oranges, which was overflowing with glistening house-tops and church-spires. Here I turned about and found a great colony of Indian Pipes.

As I turned from the shades of Wildmont, I walked toward Crystal Lake, along a dry brook bed. Here, indeed, I found a Cardinal show; over a hundred spikes of that brilliant flower danced before my eyes and lighted up the glooms. I had never before seen such flowers as these. The Cardinal-Flower (*Lobelia cardinalis*) is not frequent in Hoosac Valley—at least I have never collected it there. John Burroughs writes of it: "It is not so much something colored as it is color itself."[1] I gathered many spikes of this flaring colored flower, and passed out to the shore of the lake; children, with their sailboats, ran teasingly after me, until I escaped to a quiet retreat where ice-cream was served. The waiter and the children alike were strangely unfamiliar with this flower, growing so close to their homes.

I passed out over the rocky slopes northward, where I ate huckleberries to my heart's content. The ghost-like Feathery Plumes, and common Purple Clubs of this region towered everywhere among the woods; and low beautiful plants of the Yellow Gerardia were in full bloom. As I rounded the slope, below the Rock, I collected a fine specimen of the gorgeously colored Orange Butterfly-Weed, or Pleurisy-Root (*Asclepias tuberosa*), of the Milkweed Family. In the swamp farther south on the Orange Mountains, I have formerly collected the Swamp Milkweed flowers, which are similar to Butterfly-Weed, save that they are of a delicate rose-purple color. Our common species northward is the Purple-Flowered Silkweed. It grows along our roadside walls and river banks, and its tender leaves are used as greens, proving very delicious food.

I sat some time on the hillside under Eagle Rock, recalling the various flowers collected along the Northfield Road the year past. Llewellyn and Hutton Parks, along these summits, are always fragrant with blossoms in May and June. I once spent a holiday in Pleasant Valley beyond St. Cloud, in May and June, collecting among other

The Star-blossoms of the Grass of Paranassus (*Paranassia Carolinia*), and the Ladies' Tresses.

flowers the beautiful Tulip-Tree blossoms (*Liriodendron Tulipifera*), which some lads graciously gathered for me.

The swamps and woods about this vale produce about the same species of flowers and trees as the hills of Mosholu and Lowerre above New York City—marsh marigolds, violets, anemones, dogwoods, and glowing apple orchards that one does not soon forget. One rare flower, however, graces the Orlando Williams Swamp in Pleasant Valley that I find nowhere else. It is the Painted-Cup (*Castilleja coccinea*) of the Figwort Family. It is very similar to the Scarlet Painted-Cup that Bryant wrote about as growing on the prairies.[2]

Frequently country folk call this flower Indian's Paint-Brush; it somewhat resembles a clover tuft daubed with vermilion. The species found in New Jersey and Staten Island are the same. Thoreau found the scarlet tufts of the Painted-Cup "very common in the meadows" on Staten Island in 1843.[3] The Alpine Painted-Cups of the White and Green Mountains are somewhat different from the species found southward and westward. A friend collected flowers of these strange plants near Woodmont in the vicinity of New Haven, and about Marbledale, Connecticut. These are typical little Figworts.

The lobelias, gerardias, milkweeds, butter-and-eggs, Leopard's-Bane (*Arnica acaulis*), and field daisies are common in the pastures and woods of St. Cloud and Pleasant Valley. In the distant swamps the Sweet Bay Magnolia (*Magnolia Virginiana*) and the Tulip Tree are the only two common northern species of the Magnolia Family. A single tulip tree is found in the Hoosac Valley, at North Pownal. Tulip trees are abundant in New Haven, Connecticut, and in Bronx Park, and also on Orange Mountains. They thrive especially westward and southward, where they become beautiful flowering trees—often one hundred and forty feet high.

As I came down the Northfield Road from St. Cloud, in June, 1896, I found the pastures full of blooming briar-roses, and the meadows waving with white daisies and golden arnica. The latter flower is replaced in the meadows of the Hoosac Highlands by great patches of the Devil's Paint-Brush or Orange Hawkweed (*Hieracium aurantiacum*), an emigrant weed from Europe, which is very pretty and fragrant. The Purple Gerardia (*Gerardia purpurea*), the Blue Lobelias (*Lobelia syphilitica*), and *Lobelia spicata* grow abundantly in Pownal-on-the-Hoosac in June.

As I passed homeward through the Salt Meadows, beyond Newark, on the new Plank Road to Desbrosses Ferry, I began to observe the large pink-purple blossoms of the Swamp Rose-Mallow (*Hibiscus Moscheutos*) and the Marsh-Mallow (*Althæa officinalis*), whose roots contain a mucilaginous substance, and which are closely allied to our cultivated hollyhocks. I soon neared an open ditch by the road, filled with blossoming Arrow-Head (*Sagittaria latifolia*) and Pickerel-Weed (*Pontederia cordata*). The former produces beautiful waxen white flowers, and the latter, blue spikes of ragged blossoms. Not far from this mud-hole on the dry, sandy roadside, I gathered the rank-scented Jimson-Weed or Thorn-Apple (*Datura Stramonium*), a poisonous emigrant weed from Asia, whose Arabic name was *Tatorah*. It is common everywhere about these regions in waste ground, as well as along Kings-bridge Road and Old East Chester near the City. I have also observed it near the poor-house in New Haven, but never in the Hoosac Valley region.

The Salt Meadows of New Jersey, during August and September, are rolling swales of tall sedges and cat-tail grasses. Later in the season, when the golden-rod and purple asters are frozen and brown, and thrown in heaps upon the ground by the autumn winds, one may see great flocks of geese, and the comical purple grackle—the crow blackbird—flying southward over these desolate lands. A deep, weird solitude surrounds these unfathomable swamps. The foot of man and his bog-hoe as yet have never penetrated their regions, although within hearing of Old Trinity's chimes.

In the Hoosac Valley autumn is a season of glory. Late August produces the gorgeous colored tiger lilies. The swampy meadows in September are brightened with the delicate greenish-white stars of the Grass-of-Parnassus (*Parnassia Caroliniana*), first found on that ancient Mount Parnassus in Greece, and described and named by Dioscorides in Christ's day. Innumerable asters and golden-rod brighten the roadside hedges. In the open clearings of bushy pastures grows the Woolly Moonshine—the "everlasting" of which Thoreau wrote. It is sometimes called Cud-Weed, or Balsam-Weed (*Gnaphalium decurrens*). The Pearly-Everlasting or None-so-Pretty (*Anaphalis margaritacea*) is peculiarly fragrant and beautiful, banked in among the late golden-rods, and the crimson and chrome-colored autumn leaves of sumach and blackberry briars against the dark green pines. I have found these flowers unfolding amid the snows as late as December. Late spikes of Orchids, the Ladies' Tresses of genus *Gyrostachys*, the Bitter-Buttons or Tansy-Weed (*Tanacetum vulgare*), numerous thistles (*Carduus*), the velvety leaves of St. Peter's Mullen (*Verbascum Thapsus*), Wormwood (*Artemisia Absinthium*) grow along the roadsides over Mount Œta, while Thimble-Berry blossoms and the Bluebells-of-New-England fill in the waste places of fences and dugaway ledges.

When the cooler days of October come, we may look for that blue flower of heaven, the Fringed Gentian (*Gentiana crinita*), along the roadsides near the swamps of Etchowog, modestly and patiently waiting for the autumnal skies of blue:

Then doth thy sweet and quiet eye
Look through its fringes to the sky,
Blue—blue—as if that sky let fall
A flower from its cerulean well.[4]

The Hoosac River, Pownal, Vermont.

So come and fade alike the rarest flowers and the commonest weeds among the Highlands of the Hoosac, the valley of peaceful waters.

It is in the deepest and most secluded swamps that the shy orchid blooms, far beyond the realm of lawn or garden. Few indeed realize what a world of beauty and order lies sleeping unsought and unseen in the mossy recesses of our mountains,—a wonderland of discovery to any one who persistently, though reverently, seeks to lure from Nature the secrets of her deep retreats.

Notes:

[1] Burroughs, *Riverby.*
[2] Bryant, *The Painted Cup.*
[3] Thoreau, *Letters.* To Sophia Thoreau, May 22, 1843.
[4] Bryant, *To the Fringed Gentian.*

Flow on, fair Hoosac, with your gentle song,
Flow peacefully through all the centuries long;
To that unbounded sea, Eternity!
That God decrees alike for Man and Thee.

G. G. N.

New England Orchids

Nature, in fact that parent of all things, has produced no animated being for the purpose solely of eating; she has willed that it should be born to satisfy the wants of others, and in its very vitals has implanted medicaments conducive to health. ... Cato has recommended that flowers for making chaplets should be cultivated in the gardens: varieties remarkable for delicacy, which it is quite impossible to express, inasmuch as no individual can find such faculties for describing them as Nature does, for bestowing on them their numerous tints. Nature, who here in especial shows herself in a sportive mood, takes a delight in the prolific display of her varied productions. The other plants she has produced for our uses and our nutriment, and to them accordingly she has granted years, and even ages, of duration; but as for the flowers and their perfumes, she has given them birth for but a day—a mighty lesson to man, we see, to teach him that that which in its career is most beauteous and most attractive to the eye is the very first to fade and die.

Even the limner's art possesses no resources for reproducing colors of the flowers in all their varied tints and combinations, whether we view them in groups alternately blending their hues or whether arranged in festoons, each variety by itself.—Pliny, *Natural History* (23-79 A.D.).

Appendix

Orchidaceæ
Orchid Family

[In compiling the appendix of New England Orchids, the author has followed the order of classification and nomenclature adopted by Messrs. Britton and Brown in the *Illustrated Flora* of Northeastern North America, 1896, without doubt the highest and most systematic arrangement according to the progress of evolution and the advancement of the science of botany in North America.]

Orchidaceæ, Lindley, *Natural System*, 2d ed., p. 336. 1836.

Perennial plants arising from bulbs, corms, fibrous, or tuberous roots. Stems or scapes 2 inches to 3-4 feet high. Leaves parallel-veined, sheathing, and plicate, sometimes reduced to scales. Flowers perfect or irregular, solitary or in a spiked raceme, usually subtended by a leafy bract. Perianth consists of 6 segments. The calyx, or outer whorl, consisting of 3 parts (sepals); the corolla, or inner whorl, consisting of 3 parts (petals). The third petal is designated labellum (lip), or nectary, and is in orchids the most beautiful part, assuming grotesque shapes ornamented with spurs and fringes. The stamens and pistils are variously united with the style, forming an unsymmetrical column. Anther, 1, or in *Cypripedium* 2; 2-celled. Pollen in 2-8 pear-shaped, usually stalked masses (*pollinia*), united by elastic threads, the masses waxy or powdery, attached at the base to a viscid disk (gland). Stigma, a viscid surface, facing the labellum beneath the rostellum, or in a cavity between the anther-sacs (clinandrium). Seed-capsule (ovary) inferior, long and twisted, 3-angled, 1-celled. Ovules minute, spindle-shaped, and numerous; embryo fleshy. The colors of orchids are various and beautiful; their fragrance heavy and exquisite in several species. Orchids were known and designated by Linnaeus in 1753 as *Gynandrous*, meaning "stamens and pistils united to the column."

There are about 410 genera and from 6,000 to 10,000 species, widely distributed throughout the damp and wooded regions of the world. More abundant in the humid atmosphere of the tropics, where many species are air-plants or epiphytes. The orchids of the temperate and sub-arctic regions are terrestrial, drawing their nourishment from the earth.

North American Orchids, North of Mexico.....	150-160
New England Orchids	48-56
Hoosac Valley Orchids	40-42

Genera of Orchid Family in New England
Genera XV. Species 56.

I. *Cypripedium* Linnaeus, 1753—6 species.
II. *Orchis* Linnaeus, 1753—2 species.
III. *Habenaria* Willdenow, 1805—18 species.
IV. *Pogonia* Jussieu, 1789—4 species.
V. *Arethusa* Linnaeus, 1753—1 species.
VI. *Gyrostachys* Persoon, 1807—6 species.
  * (*Spiranthes* Richard, 1818.)
VII. *Listera* R. Brown, 1813—3 species.
VIII. *Peramium* Salisbury, 1812—4 species.
  * (*Goodyera* R. Brown, 1813.)
IX. *Achroanthes* Rafinesque, 1808—2 species.
  * (*Microstylis* Nuttall, 1818.)
X. *Leptorchis* Thouars, 1808—2 species.
  * (*Liparis* Richard, 1818.)
XI. *Calypso* Salisbury, 1807—1 species.†
XII. *Corallorhiza* R. Brown, 1813—4 species.
XIII. *Tipularia* Nuttall, 1818—1 species.†
XIV. *Limodorum* Linnaeus, 1753—1 species.
  * (*Calopogon* R. Brown, 1813.)
XV. *Aplectrum* Nuttall, 1818—1 species.†

I

CYPRIPEDIUM

Linnaeus, 1753
Lady's Slipper—Moccasin-Flower
English—Lady's Slipper.
Latin—Calceolus D. Mariae, or Marianus.

---

* Former generic designations, now antedated.
† Genera not reported for Hoosac Valley region, although native of Vermont.

The Fragrant White Moccasin-Flower. (*Cypripedium Montanum.*)

This species is a native of the Rocky Mountain region, and is closely related to our eastern Fragrant Yellow Lady's Slipper (*Cypripedium parviflorum*); these two Cypripediums being the only really fragrant species on the continent.

German—Frauenschuh, or Marienschuh.
French—Sabot de is Vierge, or Soulier de Notre Dame.
Italian—Pontoffala, or scarpa della Madonna.
Algonquin Indian—Mawcahsun, or Makkasin-Flower.
North American—Indian Moccasin-Flower.

The generic name, *Cypripedium*, comes from the Greek, referring to Κυκρισ, a former name of Venus, the Divine Mother of the Romans before Christ, and ποδιον, signifying sock, or slipper.

Glandular pubescent plants. Anthers, 2. Labellum shoe-shaped, or conical. Sepals and petals similar in texture; lower sepals wholly or imperfectly united in all species save *C. arietinum* R. Brown. Stem, 6 inches to 3 feet high. Flowers, 1-4 in the Atlantic region and 1-12 in the Pacific region; pendulous, alternating in a bracted raceme. Pollen granulose, without caudicle or glands. Fragrance heavy, aromatic, or oily save in two exquisitely sweet species, *C. parviflorum* Salisbury, of the Atlantic region, and *C. Montanum* Douglas, of the Pacific slope. Leaves, 2-several, plicate, light green, somewhat resembling *Hellebore* foliage, with which plants the *Cypripediums* were early confused by the ancient herbalists. Roots fleshy, fibrous, with spicy, oily, or musk-like odor, used as a nervine. Seed-capsule long, three-angled; ovules numerous, minute, resembling saw-dust. Seedlings frequent in many stations. About 50 species for the world.

Continental Range—Throughout the *conifer* wooded and bogland regions from Alaska southward to Mexico. May-July.

North American species	13
New England species	6
Hoosac Valley species	5

New England species

1. *C. arietinum* R. Brown, 1813.
2. *C. reginae* Walter, 1788.
   (*C. spectabile* Salisbury, 1791.)
3. *C. candidum* Willdenow, 1805.
4. *C. hirsutum* Miller, 1768.
   (*C. pubescens* Willdenow, 1805.)
5. *C. parviflorum* Salisbury, 1791.
6. *C. acaule* Aiton, 1789.

1.—*Cypripedium arietinum* R. Brown, 1813
Ram's-Head Lady's Slipper—Ram's-Head Moccasin-Flower

The specific name, *arietinum*, refers to the conical labellum resembling a ram's head.

Small *conifer* bogland or damp woodland orchid, with fibrous roots. Rare. May 9th-August 1st.

Flowers, one, terminal, mottled dull purple and white. Labellum conical, 1/2-2/3 inch, prolonged at the apex into a reflexed spur. Sepals all free. Petals narrow, similar in color, and assuming the place of horns to the ram's-head-shaped flower. Stem leafy, 6-12 inches high. Leaves, 3-4, dark apple-green; 2-4 inches wide, smooth, without hairs. Seed-capsule prominently ridged.

Continental Range—From Quebec, Ontario, southward to North Haven, Connecticut, and Mt. Toby, Massachusetts; westward to Minnesota, the Great Lake region being the centre of distribution. Limited between the 40th-50th parallels.

New England Range—Maine, rather abundant; New Hampshire, rare; Vermont, abundant; Massachusetts, rare; Connecticut, very rare.

2.—*Cypripedium reginae* Walter, 1788
(*Cypripedium spectabile* Salisbury, 1791)
White-Petaled, or Showy Lady's Slipper—
Queen of the Moccasin-Flowers

The specific name, *reginae*, refers to the queenly appearance of the white-petaled flowers.

Large bogland orchid, with fleshy fibrous roots. Frequent. June 15th-July 4th.

Flowers, 1-4 terminal, large, showy, white, tinged with deep pink or wine; the most beautiful species among our native *Cypripediums*. Labellum shoe-shaped, inflated, drooping margins of the orifice inflected, crest deeply tinged with pink-purple; interior downy, ornamented with lines of deeper purple. Rarely pure white flowers occur. Sepals and petals similar, pure white; 2 lower sepals wholly united; side petals narrower than sepals. Stem, 1-3 feet high. Leaves, 5-7, alternating to top of stem; 3-7 inches long, clasping, 1-4 inches wide; 10-13 plaits; strongly pubescent, produces poisonous effect to susceptible people similar to that caused by *Rhus*.

Continental Range—From Nova Scotia southward to the higher mountains of North Carolina, and Huntsville, Alabama; westward to Minnesota, Walhalla Mountains, North Dakota, and the Barrens of Kentucky.

New England Range—Maine, frequent; New Hampshire, frequent; Vermont, common; Massachusetts, common; Rhode Island, no stations reported; Connecticut, frequent.

The Showy Moccasin-Flower. (*Cypripedium reginæ.*)

This is the most gorgeous *Cypripedium* in the world, and without doubt one of the most ancient types of the genus. The stigma is distinctly three-lobed. The plate shows the waxy texture of the white sepals and the wine-colored crest of the shoe-shaped labellum, as well as the highly decorated interior.

3.—*Cypripedium candidum* Willdenow, 1805†
Small White Lady's Slipper—The Prairie Moccasin-Flower

The specific name, *candidum*, refers to the white labellum of this species.

Small, damp swamp-land orchid, with fleshy-fibrous roots. Rare. May 11th-June 29th.

Flowers small, solitary, and terminal. Labellum shoe-shaped, white, striped with purple interiorly; about 1 inch long; orifice small, with edges inflected. Sepals and petals lanceolate, greenish-brown and purple; lower sepals imperfectly united. Stem leafy, 6-12 inches high, pubescent. Leaves, 3-4, sheathing, erect, crowded, acute, 3-5 inches long, 2/3-1 1/2 inches wide, several scales below, 7-9 nerved.

Continental Range—From Connecticut, Pennsylvania westward to Indiana, North Dakota, Columbia Plains, on the Canadian shore of St. Clair River, to the Barrens of northern Kentucky, and Fort Hill, California, which station appears doubtful; more central distribution being from central New York to North Dakota.

New England Range—Recently reported for Connecticut by Mr. A. W. Driggs, of East Hartford.

4.—*Cypripedium hirsutum* Miller, 1768
(*Cypripedium pubescens* Willdenow, 1805)
Large Yellow Lady's Slipper—Downy Yellow Moccasin-Flower

The specific name, *hirsutum*, refers to the whole plant being hirsute, or clothed with hairs.

Large bogland or damp mountainside orchid, with fleshy-fibrous roots. May 19th-June 15th.

Flowers dull chrome yellow, 1-3, terminal, shoe-shaped, 1-2 1/2 inches long. Labellum shoe-shaped, inflated, convex above, chrome yellow, edges of orifice inflected, lined with downy hairs and dotted lines of carmine. Sepals and petals graceful, petals very much twisted; lower sepals imperfectly united; siskin-green and brown-purple. Stem leafy to top, 1-2 1/2 feet high, pubescent. Leaves, usually 5, broadly ovate, 3-5 inches long, 1 1/2-3 inches wide; 7-9 nerved; plicate and hirsute, said to cause poisonous irritation similar to *Rhus*. Roots used as a nervine.*

---

† Species not reported for Hoosac Valley region, although reported for Connecticut.

* Nicholson's *Ill. Dict. and Gard. Ency. Hort. Gard. Kew*, 1887.

Continental Range—From the wooded country of the sub-arctic lands in latitude 54°-64° North, southward throughout Canada, New England, to Alabama; westward to North Dakota, Colorado, slightly beyond the Continental Divide in the Rocky Mountain region.

New England Range—Maine, common; New Hampshire, common; Vermont, frequent; Massachusetts, abundant; Rhode Island, rare; Connecticut, rare.

5.—*Cypripedium parviflorum* Salisbury, 1791
Small Yellow Lady's Slipper—Fragrant Yellow Moccasin-Flower

The specific name, *parviflorum*, refers to the small flower of this species.

Small bogland or damp hillside orchid, with fleshy-fibrous roots. May 19th-July 4th.

Flowers small, yellow, solitary, and terminal Often intergrades with larger yellow species—(*C. hirsutum*); fragrant, the only *Cypripedium* in the Atlantic region especially so. Labellum small, 1/2-1 1/2 inch long, shoe-shaped, drooping lemon-yellow, lined with downy hairs and dotted lines of carmine. Sepals and petals brownish-purple, similar in texture; sepals 2 inches long, graceful, twisted, lower ones imperfectly united; petals glossy and twisting exceedingly. Stem leafy, slender, pubescent, 1-2 feet high. Leaves lanceolate, 3-5 inches long and 1-2 1/4 inches wide, pubescent, 7-9 nerved; not so villose as *C. hirsutum* Miller. Said to produce poisonous effect similar to *Rhus*.

Continental Range—In company with other species of New England *Cypripedium*; from Newfoundland, British Columbia, southward to Georgia; westward to the sub-humid regions of Kansas, extending slightly over the Continental Divide in Rocky Mountain region. Ascends 4000 feet altitude in Virginia.

New England Range—Maine, frequent; New Hampshire, infrequent; Vermont, frequent; Massachusetts, frequent; Rhode Island, not reported; Connecticut, rather rare.

6.—*Cypripedium acaule* Aiton, 1789
Two-Leaved Lady's Slipper—Stemless Pink Moccasin-Flower

The specific name, *acaule*, refers to the lowly and humble (acaulescent), growth of the species, since the flower is stemless, arising from a short or subterranean stem.

Sphagnous bogland, *conifer* or mixed woodland orchid with fleshy-fibrous roots. May 19th-June 20th.

The Pink Moccasin-Flower—The Stemless Lady's-Slipper.
(*Cypripedium acaule.*)

Showing the structure of the pendulous and bi-lobed labellum, and the processes of the sepals and petals. The lower sepals are wholly united in this species, and less grace or undulation appears than in the Yellow Cypripediums.

Flowers large, pink-purple, solitary, terminal, stemless. Labellum shoe-shaped, 2-2 1/2 inches long, bi-lobed, pendulous, with closed fissure down its whole length, edges inflected, downy interiorly; pink-purple with darker veining of purple. (Rarely pure white flowers occur, with chrome yellow sepals and petals.) Sepals and petals brown-purple and green, shorter than labellum; two lower sepals wholly united. Stem very short, obscured by the basal leaves. Scape naked, 8-18 inches high, single-flowered, terminal. Few instances where two flowers or buds in embryo have occurred. Leaves 2, sheathing the base of peduncle, oblanceolate 3-5 nerved, hirsute and thickened; 6-8 inches long, 2-3 inches wide, resembling the leaves of *Orchis spectabilis* Linnaeus.

Continental Range—From Newfoundland, and Fort Franklin, In latitude 54°-64° North; southward to Lookout Mountain, Mentone, and Cullman, Alabama; westward to northern Indiana, Minnesota, and Kentucky.

New England Range—Maine, common; New Hampshire, common; Vermont, abundant; Massachusetts, abundant; Rhode Island, common; Connecticut, common.

II
ORCHIS

Linnaeus, 1753
Showy Orchis

The generic name, *Orchis*, refers to orcis, the son of a rural deity of classical mythology, in whose memory these flowers were designated.

Plants with biennial roots. Anther 1. Labellum connate with base of the column; produced below into a spur. Sepals separate, free to the base, similar in texture to the petals. Flowers small, delicate white and rose-purple, fragrant; in a short terminal spike. Anther-sacs divergent; pollinia granulose, 1 in each anther-sac, which is attached at the base to a viscid disk or gland. Glands *enclosed* in a pouch. Stem, scape-like, 5-angled, 4-12 inches high. Leaves 2, oblong-obovate, shining, basal, with several bracts above. Roots fleshy-fibrous or tuberous.

Continental Range—There are three reported species of this genus for the Continent. *Orchis*, as a genus, contains 80 or more species ranging throughout the temperate regions of Europe, Asia, Northern Africa, Canaries, and North America.

North American species 3
New England species 2
Hoosac Valley species 1

The Showy Orchis. (*Orchis spectabilis.*)

The finest orchis of the season, showing the hooded fold above the orifice of the spur and the processes of the flowers on the bracted scape.

New England species

1. *O. spectabilis* Linnaeus, 1753.
2. *O. rotundifolia* (Pursh) Lindley, 1814-1835.

1.—*Orchis spectabilis* Linnaeus, 1753
Showy Orchis

The specific name, *spectabilis*, refers to the beautiful spectacle of a group of these plants in bloom.

Small, damp woodland orchid with fleshy-fibrous roots. April 19th-June 19th.

Flowers fragrant, about 1 inch long, violet-purple mined with rose-purple and white; 3-6 flowered in a bracted raceme. Labellum divergent, attached to a spur, purple and white. Sepals and petals arching in a galea. Glands *enclosed* in a pouch or hooded fold. Stem, scape-like, 4-12 inches high, thick, 5-angled. Leaves 2, basal with 1-2 scales below, and foliaceous bracts above sheathing the seed-capsules.

Continental Range—From New Brunswick, Ontario, southward to Georgia and Alabama; westward to the Rocky Mountains. Ascends 4000 feet altitude in Virginia.

New England Range—Maine, rare; New Hampshire, frequent; Vermont, frequent; Massachusetts, frequent; Rhode Island, rare; Connecticut, common.

2.—*Orchis rotundifolia* (Pursh) Lindley, 1814-1835†
Small Round-Leaved Orchis

The specific name, *rotundifolia*, refers to the round leaf of this species.

Small woodland or sphagnous bogland orchid with fleshy-fibrous roots. June 10th-July.

Flowers white, rose-purple, flecked with deeper purple, 1/2-2/3 inch long, subtended by bracts; raceme 2-6 flowered. Labellum 3-lobed, white, purple-spotted, longer than petals, central lobe largest, two-lobed or notched at the summit; spur slender, shorter than labellum. Sepals and petals oval, rose-color. Glands *enclosed*. Stem slender, 8-10 inches high. Leaf 1 near the base, orbicular or oval, 1 1/2-3 inches long, and 1-2 inches wide, sheathing scales below.

---

† Species not reported for Hoosac Valley region, although native of Vermont.

Continental Range—Rare; from Greenland, southward throughout Canada, in latitude 55°-56° North in British Columbia, Rocky Mountain region to Bristol Swamps, Addison County, Vermont, and Norfolk, Connecticut (?).

New England Range—Maine, rare; New Hampshire, rare; Vermont, rare; Massachusetts, not reported; Rhode Island, not reported; Connecticut, doubtfully reported.

III

Habenaria

Willdenow, 1805
Rein Orchis—Naked Gland Orchis

The generic name, *Habenaria*, comes from *habena*, a thong or rein.

Leafy-stemmed plants. Anther 1. Glands *naked.* Labellum spreading or drooping, with a spur at base. Sepals and petals free, similar in structure and color. Anther-sacs parallel; pollinia without caudicles, powdery or granulose. Flowers mostly in a spiked raceme, various in colors, ornamented with spurs, fringed petals and throats. Fragrance delicate and exquisite in several species. Leaves 1-many, lanceolate, becoming bract-like above. In two species—*H. orbiculata* (Pursh) Torrey and *H. Hookeriana* A. Gray—the leaves are orbicular and basal, with or without bracts above. Roots thick, fibrous, tuberous or palmate. Seedlings appear numerous in many stations.

Continental Range—A genus containing about 450-500 species for the world, widely distributed in temperate and tropical regions in Europe, Asia, and America. There are 50-55 species on the continent of North America north of Mexico.

North American species north of Mexico	50-55
New England species	18
Hoosac Valley species	13-16

New England species

1. *H. orbiculata* (Pursh) Torrey, 1814-1826.
2. *H. Hookeriana* A. Gray, 1836.
3. *H. oblongifolia* (Paine) Niles, 1865-1903.
4. *H. obtusata* (Pursh) Richardson, 1814-1823.
5. *H. hyperborea* (Linnaeus) R. Brown, 1767-1813.
6. *H. media* (Rydberg) Niles, 1901-1903.
7. *H. dilatata* (Pursh) Hooker, 1814-1825.
8. *H. fragrans* (Rydberg) Niles, 1901-1903.
9. *H. bracteata* (Willdenow) R. Brown, 1805-1813.
10. *H. clavellata* (Michaux) Sprengel, 1803-1826.

A Group of Three Species of Genus *Habenaria*.

1. The Tall Northern Green Orchis. (*Habenaria hyperborea*.)
2. The Tall Northern White Orchis. (*Habenaria dilatata*.)
3. The Large Round-Leaved Orchis. (*Habenaria orbiculata*.)

11. *H. flava* (Linnaeus) A. Gray, 1753-1840.
12. *H. ciliaris* (Linnaeus) R. Brown, 1753-1813.
13. *H. blephariglottis* (Willdenow) Torrey, 1805-1826.
14. *H. holopetala* (Lindley) A. Gray, 1835-1867.
15. *H. lacera* (Michaux) R. Brown, 1803-1810.
16. *H. grandiflora* (Bigelow) Torrey, 1824-1826.
17. *H. psycodes* (Linnaeus) A. Gray, 1753-1840.
18. *H. Andrewseii* White n. sp. (per letter, 1903).

1.—*Habenaria orbiculata* (Push) Torrey, 1814-1826
Large Round-Leaved Orchis—Heal-All—Shin-Plasters

The specific name, *orbiculata*, refers to the round or orbicular leaves of this plant.

Tall spiked woodland orchid, with thick fibrous roots. June 17th-July 15th-August 5th.

Flowers greenish-white, many in spiked raceme. Labellum oblong-linear, white, spur long. Sepals and petals 1/3 to 1/2 as long as labellum. Sepals spreading, petals smaller. Stem or scape 1-2 1/2 feet high, *bracted*, occasionally producing one small stem-leaf. Leaves 2, basal, large, round, flat-lying, 4-7 inches in diameter.

Continental Range—Not uncommon, but scarcely abundant, from Newfoundland, British Columbia, Lake Superior, southward to the western mountains of North Carolina; westward to Montana, Idaho, and Washington.

New England Range—Maine, common; New Hampshire, common; Vermont, frequent; Massachusetts, occasional; Rhode Island, not reported; Connecticut, rather rare.

2.—*Habenaria Hookeriana* A. Gray, 1836
Small Round-Leaved Orchis—Hooker's Orchis

The specific name, *Hookeriana*, refers to Sir J. Hooker, who studied this orchid and in whose honor Dr. Gray designated it.

Damp, hilly woodland orchid with fleshy-fibrous roots. June 10th-August 22d.

Flowers many, subtended by small bracts in spiked raceme, yellowish-green; spike 4-8 inches long. Labellum linear-lanceolate, acute 1/3-1/2 inch long. Sepals and petals greenish, spreading; petals awl-shaped 1/3 inch long. Stem 8-18 inches high, *not bracted*. Leaves 2, oval, obovate, or orbicular, slightly ascending, 3-6 inches long.

Continental Range—From Nova Scotia, Lake Huron, Lake Superior, southward to New Jersey and Pennsylvania, westward to Indiana, Minnesota, Wisconsin, and Iowa.

New England Range—Maine, common; New Hampshire, frequent; Vermont, frequent; Massachusetts, occasional; Rhode Island, rare; Connecticut, rare.

3.—*Habenaria oblongifolia* (Paine) Niles, 1865-1903
Small Oblong-Leaved Orchis

The specific name, *oblongifolia*, refers to the oblong leaves.

Damp, hilly woodland orchid with fleshy-fibrous roots. June-August.

Flowers many in spiked raceme, yellowish-green similar or identical with the spike of flowers of *H. Hookeriana*. Stem 8-18 inches high, *not bracted*. Leaves 2, *oblong*, ascending.

Continental Range—In similar situations with *H. Hookeriana* and *H. orbiculata*, from Nova Scotia (Macoun), Campbellton, New Brunswick, (Chalmers), Chelsea Mountains, Quebec (Fletcher's *Flora*, Ottawa); southward to New York (Paine and Dudley), New Jersey (Mrs. Britton), and throughout New England States; westward to Iowa.

4.—*Habenaria obtusata* (Pursh) Richardson, 1814-1823†
Sub-Alpine Greenish Bog-Orchis

The specific name, *obtusata*, refers to the obtuse or blunt sepals and labellum of this species.

Small sub-alpine bogland orchid, with fibrous roots. June 22d-July 30th-September.

Flowers, greenish-yellow in loose spiked raceme 1-2 1/2 inches long; flowers 1/4 inch long. Labellum blunt or obtuse, deflexed, entire. Lateral sepals spreading, oblong and obtuse. Petals shorter, obtusely 2-lobed at base. Stem slender, *not bracted*, 4-10 inches high, 4-angled. Leaf 1, basal, obovate.

Continental Range—From Alaska, southward throughout Canada to Mt. Wachusett, Massachusetts; westward to Minnesota, Wyoming, Montana, and Colorado.

---

† Doubtfully reported for Hoosac Valley region, although native of Vermont.

New England Range—Maine, frequent; New Hampshire, frequent; Vermont, rare; Massachusetts, very rare; Mt. Wachusett (Dr. G. E. Stone); Mt. Washington (Henry Baldwin); Rhode Island, not reported; Connecticut, not reported.

5.—*Habenaria hyperborea* (Linnaeus) R. Brown, 1767-1813
Tall Green Northern Orchis

The specific name, *hyberborea*, refers to the species being tall and a boreal or northern orchid.

Cold bogland or damp woodland orchid, with thick fleshy roots. May 30th-July 28th-August 18th.

Flowers small, greenish-yellow, on bracted spike 3-8 inches long; infrequent, said to be constructed for self-fertilization, if insects fail to visit the flowers. Labellum lanceolate, obtuse, and entire. Sepals and petals obtuse, ovate, 1/6-1/4 inch long; upper sepal crenulate at apex. Stem tall, stout, leafy, 8 inches to 3 feet high. Leaves many lanceolate, acute. Seed-capsule much twisted.

Continental Range—From Greenland, Yakutat Bay and eastern part of Kodiak Island and vicinity of Sitka, Alaska, to Fort Franklin; southward to Pennsylvania and New Jersey; westward to Minnesota, Montana, Washington, New Mexico, and California. This species is closely allied with the Tall White Fragrant Bog-Orchis (*H. dilatata*). The latter, however, is not constructed for self-fertilization. Several species of *Habenaria* appear to intergrade with each other more or less throughout their range.

New England Range—Maine, common; New Hampshire, frequent; Vermont, abundant; Massachusetts, infrequent; Rhode Island, not reported; Connecticut, rare.

6.—*Habenaria media* (Rydberg) Niles, 1901-1903†
Intermediate Bog-Orchis

The specific name, *media*, refers to the intermediate form of this species between *H. hyperborea* and *H. dilatata*.

A tall bogland orchid, with fleshy roots. June-August.

† Doubtfully reported for Hoosac Valley region, but should be looked for wherever *H. hyperborea* grows.

The Spikes of Habenaria.
(*Habenaria Andrewseii* and *Habenaria psycodes.*)

Flowers greenish-purplish, spike densely flowered; it has often been confused with *H. hyperborea* and *H. dilatata*, which it closely resembles. Type specimen from Quebec. Labellum lanceolate, entire, obtuse, slightly dilated at the base; spur exceeding the labellum, curved and obtuse. Sepals ovate-oblong; petals lanceolate, obtuse. Stem 16 inches 2 1/2 feet high, rather stout. Leaves lanceolate, acute. Seed-capsule 2/5 inch long

Continental Range—In bogs from Quebec, southward to New York and New England.

New England Range—There are no authoritative stations reported as yet, although the author collected intermediate forms of *H. hyperborea* in Dimmick Swamp, Pownal, Vermont, July, 1903, answering to Dr. Rydberg's descriptions of this form. It is also reported by Marcus White and A. Le Roy Andrews for Pownal Swamps.‡

7.—*Habenaria dilatata* (Pursh) Hooker, 1814-1825
Tall White Northern Orchis

The specific name, *dilatata*, refers to the dilated condition of the labellum of this orchid.

A tall slender or stout bogland orchid, with fleshy-fibrous or tuberous roots. June 2d-August 23d.

Flowers white, small, in a densely flowered spike 2-10 inches long. Slightly fragrant. Not constructed for self-fertilization as the closely allied species *H. hyperborea*. Labellum entire, dilated, or obtusely 3-lobed at base. Spur blunt and incurved. Sepals ovate, obtuse, and small. Stem slender, often stout, inferring that an intermediate form exists, which Dr. Rydberg has designated specifically as *fragrans*. Leaves lanceolate. Seed-capsule much twisted.

Continental Range—From Ankow River, Ocean Cape, Alaska, and Unalaska in latitude 60° North; southward to Litchfield, Connecticut, Amherst, Massachusetts, and Pennsylvania; westward to Minnesota, Wyoming, Colorado, also occurring in the cañons of Clear Water Valley, Idaho.

New England Range—Maine, common; New Hampshire, frequent; Vermont, frequent; Massachusetts, infrequent; Rhode Island, not reported; Connecticut, rare.

‡ A. L. Andrews, *Rhodora*, 4: 79-82, 1902.

8—*Habenaria fragrans* (Rydberg) Niles, 1901-1903†
Fragrant Slender Bog-Orchis

The specific name, *fragrans*, refers to the exquisite fragrance of this species, which is so closely allied with *H. dilatata* and of which it appears to be a form.

Slender bogland orchid, with fleshy-fibrous roots. July.

Flowers small, pure white, very fragrant; in a slender spiked raceme. Labellum narrowly linear, dilated at the base, obtuse, shorter than the curved filiform spur; otherwise as the preceding species. Sepals lanceolate, acutish, strongly striate. Stem very slender and leafy above, 8-12 inches high. Leaves linear, several.

Continental Range—Reported from a single station in Vermont. Slender forms of *H. dilatata*, very fragrant and slightly so, are present in the bogs of Pownal, Vermont, and North Adams, Massachusetts, where the writer has collected them, believing that they were forms brought about by the intergrading of the bogland *Habenarias* closely associated. Species of *Habenaria* appear to produce natural hybrids readily.‡

New England Range—Vermont.

9.—*Habenaria bracteata* (Willdenow) R. Brown, 1805-1813
Long Bracted Orchis

The specific name, *bracteata*, refers to the long bracts, subtending the seed-capsules of this species.

A slender bracted bogland orchid, with fleshy-fibrous roots. May 8th-July 14th-August 12th.

Flowers small, greenish, in a loosely flowered spike, 3-5 inches long, subtended by long bracts. Labellum long, spatulate 2-3 toothed or lobed, twice as long as sac-like spur. Sepals ovate-lanceolate, spreading, dilated, at base; petals very narrow, thread-like. Stem leafy, slender, or stout, 6 inches to 2 feet high. Leaves lanceolate, oval.

Continental Range—From Sitka and Unalaska to the Great Plains, approaching the Rocky Mountains in latitude 55° North; southward throughout New England to North Carolina; westward to Minnesota, North and South Dakota, and Montana.

---

† Doubtfully reported for Hoosac Valley region, although native of Vermont.

‡ A. L. Andrews, *Rhodora*, 4 : 79-81, 1902.

The Small Bog. (*Habenaria clavellata.*)

New England Range—Maine, common; New Hampshire, common; Vermont, common; Massachusetts, frequent; Rhode Island, rare; Connecticut, rather rare.

10.—*Habenaria clavellata* (Michaux) Sprengel, 1803-1826
Small Yellowish Bog-Orchis—Small Wood-Orchis

The specific name, *clavellata*, refers to the club-shaped appendages of stigma or clavate spur attached to the flowers of this species.

A small bogland or woodland orchid, with fibrous roots. May 17th (Missouri), June (Alabama); July 15th-August (Maine).

Flowers small greenish-yellow, in a loosely flowered spike 1 1/2-2 inches long. Labellum dilated, 3-toothed at apex; spur longer than ovary, clavate. Sepals and petals ovate. Stem 8-18 inches high. Leaf 1, near the base, 1-3 bracts above.

Continental Range—From Lake Huron, Newfoundland, southward to Alabama; westward to Indiana, and Missouri. Ascends 6000 feet altitude in North Carolina.

New England Range—Maine, frequent; New Hampshire, frequent; Vermont, infrequent; Massachusetts, infrequent; Rhode Island, frequent; Connecticut, frequent.

11.—*Habenaria flava* (Linnaeus) A. Gray, 1753-1840
Tubercled Orchis

The specific name, *flava*, comes from the Latin *flavous*, referring to the yellow flowers of this orchid.

A damp meadow or sphagnous woodland orchid, with fleshy-fibrous roots. May (Florida), June 5th-August 25th (New England).

Flowers dull greenish-yellow, small, in a spiked raceme. Labellum entire, crenulated with obtuse tooth on each side, and a central tubercle at middle of base. Sepals and petals roundish, 1/4 inch long. Sepals yellowish. Stem stout leafy 1-2 feet high. Leaves elliptic, acute; bracts longer than seed-capsule.

Continental Range—From Crow River, Hastings County, Ontario, Canada, southward to Alabama and the wooded Tidal Swamps, Duval County, Florida; westward to Minnesota and Missouri.

New England Range—Maine, common; New Hampshire, frequent; Vermont, infrequent; Massachusetts, occasional; Rhode Island, common; Connecticut, common.

12.—*Habenaria ciliaris* (Linnaeus) R. Brown, 1753-1813†
Yellow-Fringed Orchis

The specific name, *ciliaris*, refers to the fringed labellum, from the Latin *ciliate*,—beset with *cilia* or hairs, like eyelashes fringing the eyelids.

A tall, wet, sandy meadow or pine barren orchid, with small fibrous roots. June-July 7th-August 14th.

Flowers large, showy, orange-yellow, fringed, in a many-flowered, spiked raceme, 3-6 inches long; nearly 3 inches broad. Labellum oblong and fringed. Sepals orbicular to ovate; lateral sepals reflexed; spur 1-1 1/2 inch long, very slender; petals toothed, oblong, much smaller. Stem slender, leafy, 1-2 1/2 feet high. Leaves lanceolate, 4-8 inches long, becoming bract-like above.

Continental Range—From Essex County, Canada, southward to New England, the pine barrens of New Jersey, Alabama, and Florida; westward to Illinois and Nebraska. The fringed *Habenarias* are among our most beautiful native orchids, gracing lake-side solitudes with their flame-like spikes of purple, white, or orange.

New England Range—Maine, not reported; New Hampshire, rare; Vermont, rare; Massachusetts, rare; Rhode Island, rare; Connecticut, rare.

13.—*Habenaria blephariglottis* (Willdenow) Torrey, 1805-1826
White-Fringed Orchis

The specific name, *blephariglottis*, refers to the fringed throat of this orchid.

A slender bogland orchid, in similar situations with *H. ciliaris*, with fleshy-fibrous roots. June (Alabama)-July 23d-August (New England).

Flowers pure white, in a loosely and many-flowered spiked raceme, smaller than those of *H. ciliaris*, with which it seems to intergrade. Labellum narrow, oblong, slightly fringed. Petals toothed or sparsely fringed. Stem leafy, slender, 1-2 feet high.

Continental Range—From Newfoundland southward to New England, North Carolina, and Alabama; westward to Minnesota. Intermediate forms between the Yellow-Fringed and the White-Fringed *Habenarias* are probably natural

---

† Species not reported for Hoosac Valley region, although native of Vermont.

hybrids, as they are closely associated in their haunts during July and August, the hybrid usually being of a lighter yellow, blooming slightly earlier than the type species.

New England Range—Maine, common; New Hampshire, frequent; Vermont, infrequent; Massachusetts, frequent; Rhode Island, infrequent; Connecticut, rare.

14.—*Habenaria holopetala* (Lindley) A. Gray, 1835-1867†
Cream-Fringed Orchis

The specific name, *holopetala*, refers to the petals being complete, entire.

A small bogland orchid, with fleshy-fibrous roots. July-August.

Flowers beautiful, smaller than in preceding species, cream, or lighter yellow than *H. ciliaris*. Labellum sparingly fringed. Petals narrower and entire. Stem 1 foot high. Leaves similar to type species *H. blephariglottis* and *H. ciliaris*.

Continental Range—From Canada southward to Alabama, North Carolina, in company with *H. blephariglottis*.

New England Range—Refer to range of *H. blephariglottis* for New England.

15.—*Habenaria lacera* (Michaux) R. Brown, 1803-1810
Ragged-Fringed Green Orchis

The specific name, *lacera*, refers to the lacerate or ragged and torn appearance of the fringed labellum of this species.

A bogland, meadow, or woodland orchid, with fleshy-fibrous roots. June 20th-July 15.

Flowers greenish-yellow, in loose-flowered spike, 2-6 inches long. Labellum 3-parted, deeply fringed or ragged. Sepals ovate, obtuse, upper one broader than lower ones. Stem slender or stout, leafy, 1-2 1/2 feet high. Leaves lanceolate, 5-8 inches long; bract-like above.

Continental Range—From Nova Scotia southward to Georgia and Alabama; westward to Minnesota and Indiana.

In Thompson's Swamps, Pownal, Vermont, this species intergrades with *H. psycodes* and *H. clavellata*. The hybrids, or those crossed with *H. psycodes* produce flowers with light

† Species not reported for Hoosac Valley region, although native of Vermont.

purple, and less fringed than in the true type of *H. lacera*; while those crossed with *H. clavellata* are much less ragged fringed than in the type form of *H. lacera*, although in a much smaller spike, with nearly white or creamy-green flowers, resembling a large spike of *H. clavellata*.

New England Range—Maine, common; New Hampshire, common; Vermont, common; Massachusetts, infrequent; Rhode Island, frequent; Connecticut, common.

16.—*Habenaria grandiflora* (Bigelow) Torrey, 1824-1826
(*Habenaria fimbriata* A. Gray, 1867)
Large Purple-Fringed Orchis—Long Purples—Dead-Men's Fingers—Dead-Men's Thumbs

The specific name, *grandiflora*, refers to the grand flowers of this species; the most beautiful among the bogland *Habenarias*.

Tall, damp woodland or bogland orchid, with fleshy-fibrous or palmate roots. June-July 22d-August 16th.

Flower lilac or deep purple, fragrant in densely-flowered spiked raceme; 3-15 inches long, 1-2 1/2 inches broad. Rarely white flowered spikes occur. Labellum 3-parted, 1/2-1 inch broad; 1/2 inch long, deeply fringed. Sepals and petals connivent, somewhat toothed. Spur 1-1 1/2 inch long, clavate. Stem leafy, 1-5 feet high. Leaves numerous, oblong, 4-10 inches long, 1-3 inches wide; bract-like above.

Continental Range—From Nova Scotia southward to North Carolina; westward to Wisconsin and Michigan.

New England Range—Maine, common; New Hampshire, common; Vermont, frequent; Massachusetts, frequent; Rhode Island, frequent; Connecticut, common.

17.—*Habenaria psycodes* (Linnaeus) A. Gray, 1753-1840
Small Purple-Fringed Orchis—Long Purples—Dead-Men's Fingers—Dead-Men's Thumbs

The specific name, *psycodes*, comes from the Greek *psychoda*, a butterfly, probably referring to the dainty poise of the fringed, and winged-petaled flowers.

Tall bogland or damp woodland orchid, with fleshy or palmately-tuberous roots. July 1st-August-September.

Flowers pink-purple, rarely white, very fragrant in densely flowered spiked raceme, 2-6 inches long; 1-1 1/2 inches broad. Labellum 3-parted, fan-shaped, fringed, 1/3-1/2 inch broad, much smaller and often confused with the larger species (*H. grandiflora*). Sepals and petals similar in texture, lower sepals ovate, upper one narrower. Petals oblong, toothed on upper margin. Spur clavate at apex. Stem leafy, 1-3 feet high.

Continental Range—From Newfoundland, Nova Scotia, Lake Huron southward to the swampy meadows in the western mountains of North Carolina; westward to Minnesota in tamarack sphagnous swamps.

New England Range—Maine, very common; New Hampshire common; Vermont, common; Massachusetts, frequent; Rhode Island, common; Connecticut, frequent.

18.—*Habenaria Andrewseii* White n. sp.(per letter), 1903
Andrews' Rose-Purple Orchis
Habenaria psycodes X *lacera* White and Andrews,
Rhodora 3: 245-248, 1901

The specific name, *Andrewseii*, refers to the species being named in honor of Dr. A. LeRoy Andrews, who described this species as a varietal form of *H. psycodes* X *lacera*. The species was first discovered and collected by Mr. Marcus White, July 22, 1899.

Tall bogland orchid, with fleshy roots. July 22d-August 16th.

Flowers white, to rose-purple. Labellum about 1/3-1/2 inch broad, divisions deeply cleft as in *Habenaria lacera*, yet more numerous. Sepals round-oval, obtuse. Petals cuneate-spatulate and denticulate above. Arms of column slightly acute, similar to *Habenaria psycodes*. Pollen stalked, greenish-yellow, somewhat 2-lobed, not obstructing the orifice of nectary, as observed in *Habenaria lacera*. Spur longer than ovary, clavate, much enlarged below. Ovary intermediate or rather short. Stem leafy 1-2 1/2 feet high. Raceme of flowers not as broad as in *Habenaria psycodes*, but more nearly resembling the spike of *Habenaria lacera* in form, yet fewer-flowered. Leaves several, as in *Habenaria lacera*.

Continental Range—Pownal Swamps, southwestern Vermont; no other station appears to be reported for this natural hybrid of the Purple-Fringed and Ragged Orchises (*H. psycodes* and *H. lacera*), yet it appears to be flourishing and increasing in numbers in its special haunts.

New England Range—Vermont, rare; Pownal Swamps August 5, 1901 (Marcus White and A. LeRoy Andrews); August 10-16, 1903 (Grace G. Niles).

Andrews' Rose-Purple Orchis. (*Habenaria Andrewseii.*)

IV

POGONIA

Jussieu, 1789
Sweet Pogonias

The generic name, *Pogonia*, comes from the Greek Πωγωνιασ, signifying bearded, from the bearded labellum of the type species.

Small bogland or woodland orchids, with fibrous creeping roots. Anther, 1. Labellum erect from base of the column, spurless. Sepals and petals free. Anther terminal, stalked, attached to back of column. Pollinia, 2, 1 in each anther-sac, powdery-granular, without caudicle. Flowers solitary or 1-3 in terminal raceme. Leaves 1-5, alternating, or in whorls as in *Pogonia verticillata*. Seedlings numerous in many places.

Continental Range—There are about forty-five species of this genus distributed in the wooded regions of the world. The species of *Pogonia* have formerly been confused with *Arethusa*.

North American species north of Mexico	5
New England species	4
Hoosac Valley species	3-4

New England species

1. *P. ophioglossoides* (Linnaeus) Ker, 1753-1816.
2. *P. trianthophora* (Swartz) B. S. P., 1800-1888.
3. *P. verticillata* (Willdenow) Nuttall, 1805-1818.
4. *P. affinis* Austin, 1867.

1.—*Pogonia ophioglossoides* (Linnaeus) Ker, 1753-1816
Rose Pogonia—Snake-Mouth

The specific name, *ophioglossoides*, refers to the flower resembling a snake's month.

Small sphagnous swamp-land orchid, with fibrous creeping roots. April-May (Florida), June 21st-July 20th (New England).

Flowers 1, sometimes 2, terminal, nodding, rather large for plants; rose and purplish-yellow; fragrant, subtended by a foliaceous bract. Labellum free, somewhat appressed to the column below, fringed and spurless. Sepals and petals about equal, similar. Stem 8-15 inches high. Leaves 1-3, rarely 1 basal leaf; the stem-leaf is 1/2-3 inches long, bluntly acute; foliaceous bract subtending seed-capsule.

Continental Range—From Newfoundland, southward to the pine barrens of Alabama and Florida; westward to Minnesota and Kansas.

New England Range—Maine, common: New Hampshire, common; Vermont, common; Massachusetts, common; Rhode Island, common; Connecticut, common.

2.—*Pogonia trianthophora* (Swartz) B. S. P., 1800-1888†
Nodding Pogonia

The specific name, *trianthophora*, refers to the 3-lobed labellum and the usually 3 nodding flowers of this species.

Small woodland orchid, with tuberous roots. July-August 11th-September 24th.

Flowers 1-7 on axillary peduncles, pale purple, at first nearly erect, soon drooping, 1/2-2/3 inch long. Labellum 3-lobed, clawed, roughened, without a spur. Sepals and petals equal, connivent. Stem 3-8 inches high. Leaves, 2-8, alternate, ovate, clasping stem. Seed-capsule oval, drooping.

Continental Range—From Canada, southward to North Carolina, Alabama, and Kentucky; westward to Illinois, Missouri, Iowa, and Indiana.

New England Range—Maine, rare; New Hampshire, infrequent; Vermont, rare; Massachusetts, rare; Rhode Island, rare; Connecticut, infrequent.

3.—*Pogonia verticillata* (Willdenow) Nuttall, 1805-1818
Large Whorled Pogonia

The specific name, *verticillata*, refers to the whorled or verticillate growth of the leaves.

Small moist woodland orchid, with fleshy-fibrous creeping roots. May 20th-June 15th.

Flowers solitary, purplish-green, and yellow, erect or declined, terminal; peduncle 1/2-2/3 inch long, in fruit exceeding the seed-capsule. Labellum 3-lobed, crested along a narrow band, upper part expanded, greenish-yellow. Sepals linear, 1 1/2-2 inches long, 1/12 inch wide, spreading, dark purple; petals linear, obtuse, greenish-yellow, 3/4 inch long. Stem 10-12 inches high, round and purplish. Leaves 5, verticillate, in a whorl strongly recalling the whorls of the Indian Cucumber plants (*Medeola*), with which it grows in company; obovate, pointed at apex, 1-3 inches long. Rarely an obovate

† Species doubtfully reported for the Hoosac Valley region, although native of Vermont.

The Beautiful Arethusa. (*Arethusa bulbosa.*)

Showing the structural parts of the flower, the single leaf, and bulbous root.

basal stem-leaf occurs below the whorl. The roots distinguish this plant from Indian Cucumber, yet many times the two plants so closely resemble each other that it is difficult to determine one from the other until the roots are revealed. The stem of the *Pogonia*, however, is larger, fleshy, purplish, and juicy, while the Indian Cucumber is brittle and slender.

Continental Range—Prom Ontario, Canada, southward to North Carolina and Florida; westward to Michigan and Indiana.

New England Range—Maine, not reported; New Hampshire, rare; Vermont rare; Massachusetts, frequent; Rhode Island, rare; Connecticut, frequent.

4.—*Pogonia affinis* Austin, 1867†
Small Whorled Pogonia

The specific name, *affinis*, comes from the Latin *affinitas*, near alliance, referring to the close relation or affinity this species bears to its sister species, *Pogonia verticillata*.

Small moist woodland orchid, with fleshy-fibrous roots. June.

Flowers 2 or solitary, greenish-yellow, peduncle 1/6-1/3 inch long. Labellum crested over the whole face and lobes. Sepals and petals equal. Stem 8-10 inches high. Leaves in a whorl of 5 at the summit; smaller than the whorl in *P. verticillata*. Seed-capsule erect, 1 inch long.

Continental Range—From northern Vermont, southward to Pennsylvania. This orchid has quite recently been collected in Vermont, by Mrs. Henry Holt, near Burlington (1902). Several stations in New Jersey have been destroyed since Mr. Austin first identified it there.

New England Range—Maine, not reported; New Hampshire, not reported; Vermont, very rare; Massachusetts, very rare; Connecticut, frequent; it should be more common in this State, since *P. verticillata* is frequent.

V
ARETHUSA

Linnaeus, 1753
Beautiful Arethusa

† Species not reported for Hoosac Valley region, although native of Vermont.

The generic name, *Arethusa*, refers to the dedication of this species to the nymph Arethusa of classical literature.

Small scapose orchids with bulbous roots; one of the most beautiful native orchids. Anther 1. Labellum dilated, recurved, bearded down the face, spreading at the apex. Sepals and petals similar, arched above, coherent below. Anther operculate. Pollinia 4; 2 in each anther-sac, powdery, granular. Scape 5-10 inches high, glabrous. Leaves 1, 4-6 inches long, linear, and hidden. 1-3 bracts or scales below. Seed-capsule 1 inch long, ellipsoid, 6-ribbed, rarely maturing.

Continental Range—In sphagnous bogs from Newfoundland, southward to North Carolina; westward to Minnesota. Rare throughout its range, seeking unfathomable cranberry marshes, among bushes of Kalmia and Labrador Tea, in close company with Rose Pogonia and Grass Pinks. A genus, consisting of 3 species for the world.

North American species north of Mexico	1
New England species	1
Hoosac Valley species	1

New England species

1. *A. bulbosa* Linnaeus, 1753.

1.—*Arethusa bulbosa* Linnaeus, 1753
Beautiful Arethusa

The specific name, *bulbosa*, refers to the bulbous roots of this orchid.

Small scapose, bogland orchid with bulbous roots. May 17th -June 30th.

Flowers 1, rarely 2; rose-purple, terminal, nodding, arising from between 2 unequal scales; 1-2 inches long. Labellum drooping, dilated, recurved, spreading at apex, often fringed or toothed, variegated with purple blotches and yellow. Sepals and petals linear, obtuse, arched over the column. Scape 5-10 inches high, glabrous, producing 1-3 sheathing bracts. Leaf 1, linear, many-nerved, 4-6 inches long, hidden above bracts. Root small, onion-like bulb. Seed-capsule 1 inch long, ellipsoid, 6-ribbed, rarely maturing, although seedlings appear numerous in natural haunts, the sphagnum being filled with lightly-attached bulbs.

Continental Range—From Newfoundland, southward to North Carolina; westward to Minnesota.

New England Range—Maine, common; New Hampshire, common; Vermont, common; Massachusetts frequent; Rhode Island, common; Connecticut, common.

The Hooded Ladies' Tresses. (*Gyrostachys Romanzoffiana.*)

VI
GYROSTACHYS

Persoon, 1807
(*Spiranthes* L. C. Richard, 1818)
Ladies' Tresses

The generic name, *Gyrostachys*, refers to the twisting spikes, which resemble a "coil," or "curl," from which originated the common name of Ladies' Tresses.

Erect spiked racemes of twisting flowers. Anther 1. Labellum clawed, concave, erect, embracing the column. Sepals free, coherent at top with petals, forming a galea. Flowers small, spurless, white or greenish, in rows. Anther without a lid, situated back of column. Pollinia 2, 1 in each anther-sac, powdery. Fragrance delicate. Stem leafy, bracted above basal leaves. Leaves linear, save in two species—*G. simplex* and *G. gracilis*, in which they are round, oblong, and basal. Roots fleshy-fibrous, or tuberous. Seed-capsule erect and oblong.

Continental Range—Common in pasture-land and meadow boglands; from Alaska, southward to Florida; westward to the Pacific coast. There are 80 or more species of this genus distributed throughout the tropical and temperate regions of the world.

North American species north of Mexico	19-20
New England species	6
Hoosac Valley species	5-6

New England species

1. *G. Romanzoffiana* (Chamisso), MacMillan, 1828-1892.
2. *G. plantaginea* (Rafinesque), Britton. 1818-1896.
3. *G. ochroleuca* Rydberg, 1901.
4. *G. cernua* (Linnaeus), Kuntze, 1753-1891.
5. *G. simplex* (A. Gray), Kuntze, 1867-1891.
6. *G. gracilis* (Bigelow), Kuntze, 1824-1891.

1.—*Gyrostachys Romanzoffiana* (Chamisso), MacMillan, 1828-1892
Hooded Ladies' Tresses

The specific name, *Romanzoffiana*, refers to the species being named in honor of Count Romanzov, a Russian nobleman.

Rather conspicuous bogland orchid, with 1-6 tuberous roots. July 15th-August 27th-September 8th.

Flowers white or greenish, 3-rowed, in spiked racemes, slightly twisted, 2-4 inches long, 1/3-2/3 inch thick; very fragrant. Labellum oblong, contracted below the crisped apex. Sepals and petals broad at the base, hooded above. Stem 6-15

The Nodding Ladies' Tresses. (*Gyrostachys cernua.*)

inches high. Leaves below, near the base, 3-8 inches long, with bracts above. Seedlings produce but one tuber, older plants several adhering, older tubers finally withering.

Continental Range—From Unalaska and New Metlakatla, along the southeastern coast of Alaska, southward to Pennsylvania; westward to the Pacific Ocean, ascending 8500 feet altitude in moist meadows in Yellowstone Park, Montana, and 9500 feet on Mount Graham, in Arizona. It is especially a Northern species.

New England Range—Maine, frequent; New Hampshire, infrequent; Vermont, infrequent; Massachusetts, rare; Reservoir Swamp, Notch Road, North Adams, August 17th, 1903 (Grace G. Niles); swamps, base of Mount Greylock, Adams, (Marcus White); Mr. White is the first botanist to collect this species in Massachusetts.

2.—*Gyrostachys plantaginea* (Rafinesque) Britton, 1818-1896
Early Broad-Leaved Ladies' Tresses

The specific name, *plantaginea*, refers to a supposed resemblance of this species to the spikes and foliage of certain species of Plantain (*Plantago*).

Small moist woodland orchid, with fleshy-fibrous roots. June 7th-July 2nd-August 15th.

Flowers yellow and white, spreading, 1/4 inch long, in densely flowered raceme, 1-2 inches long, 1/3-1/2 inch thick. Labellum pale yellow on face, oblong, not contracted in middle, apex rounded and fringed, base clawed. Sepals and petals white; lateral sepals free, lanceolate, upper sepal, united with petals. Stem 4-10 inches high. Leaves 4-5, oblanceolate, 1-5 inches long, becoming bract-like above.

Continental Range—From New Brunswick, Ontario, southward to Virginia; westward to Wisconsin and Minnesota. Rather rare in the southern and western range; nowhere abundant.

New England Range—Maine, rare; New Hampshire, rare; Vermont, frequent; Massachusetts, rare; Rhode Island, not reported; Connecticut, rare.

3.—*Gyrostachys ochroleuca* Rydberg, Britton's *Manual*. 300, 1901†
Yellow Ladies' Tresses

The specific name, *ochroleuca*, refers to the yellowish-green or ochroleucous color of the flowers.

---

† Species doubtfully reported for Hoosac Valley region, although native of Massachusetts.

Small hillside and damp pasture-land orchid, with fleshy or tuberous roots. August-September 1st-15th.

Flowers yellowish-green, very fragrant, in densely flowered raceme, conspicuously acute in bud, of creamy-green color, in company with *G. cernua*, but not so abundant. Labellum oblong, crenulate or crisped. Stem 12-20 inches high, pubescent. Leaves linear, tapering at both ends, basal, with bract-like leaves above. Slightly later than *G. cernua*.

Continental Range—From New Hampshire, southward to Pennsylvania, and North Carolina.

New England Range—Without doubt in each State in company with *G. cernua*, of which it appears to be a form. Frequent Mt. Washington, Massachusetts, and swamps about North Adams, September 1st (Grace G. Niles).

4.—*Gyrostachys cernua* (Linnaeus) Kuntze, 1753-1891
Nodding Ladies' Tresses

The specific name, *cernua*, refers to cernuous or nodding flowers of this species.

Small bogland or damp meadow orchid, with fleshy or tuberous roots. August 25th-September 1st-28th-October.

Flowers white, fragrant, nodding or spreading, 1/2 inch long, in three rows; densely flowered raceme, twisted, 4-5 inches long, 1/2-2/3 inch thick. Labellum oblong, apex rounded, crisped. Lateral sepals free; upper one arching with petals. Stem 6-25 inches high. Leaves at or near base, linear-oblanceolate, 3-14 inches long, with 2-6 bracts above.

Continental Range—From the Barren Grounds in latitude 64°-69° North, southward to Florida; westward to Minnesota, Kansas, Indian Territory, Texas, and New Mexico.

New England Range—Maine, common; New Hampshire, common; Vermont, common; Massachusetts, common; Connecticut, very common.

5.—*Gyrostachys simplex* (A. Gray) Kuntze, 1867-1891†
Little Simple Ladies' Tresses

The specific name, *simplex*, refers to the simplicity of this species.

Slender sandy soil orchid, with *single* tuberous root. August-September 9th.

† Species not reported for Hoosac Valley region, although native of Massachusetts.

Flowers, white, small, 1/12 inch long; in slender, twisting, spiked raceme, 1 inch long. Labellum obovate-oblong, short-clawed, and crisped at the summit; callosities nipple-shaped. Stem simple and slender, 5-9 inches high, with small bracts above basal leaves, glabrous, slightly twisted. Leaves 2-3 basal, oblong and short, disappearing at or after flowering season.

Continental Range—Front Massachusetts, southward to Delaware, New Jersey, Pennsylvania, Maryland, and Tennessee. It appears more frequent near the coast, from Nantucket, Delaware, Staten Island, and throughout New Jersey, in company with *G. gracilis*.

New England Range—Massachusetts, rare; Rhode Island, rare; Connecticut, infrequent.

6.—*Gyrostachys gracilis* (Bigelow) Kuntze, 1824-1891
Slender Ladies' Tresses

The specific name, *gracilis*, refers to the slender and grass-like spike of this species.

Small sandy woodland or pasture-land orchid, with several spindle-shaped tuberous roots. April 15th, Florida (Curtiss); May, Alabama; July 25th-September-October 15th, both North and South.

Flowers white, very fragrant, 1/6-1/4 inch long, in a slender spiked raceme, 1-3 inches long, 1/3-1/2 inch thick, loose, usually much twisted. Labellum white on margins, thick and greenish in middle, 1/6 inch long, clawed at the base, crenulate at the apex. Stem 8-24 inches high, slender, grass-like. Leaves 3 obovate, sometimes nearly orbicular, basal, dying away at or before flowering season.

Continental Range—From Nova Scotia, southward throughout Canada, and New England, to Florida; westward to Minnesota, and Texas.

New England Range—Maine, common; New Hampshire, very common; Vermont, common; Massachusetts, common; Rhode Island, very common; Connecticut, common.

VII
LISTERA

R. Brown, 1813†
Lister's Twayblade

† Genus doubtfully reported for Hoosac Valley region, although native of Vermont.

The Slender Ladies' Tresses. (*Gyrostachys gracilis.*)

Showing the spindle-shaped roots; this species is closely allied with *G. simplex*, which produces but a single tuber.

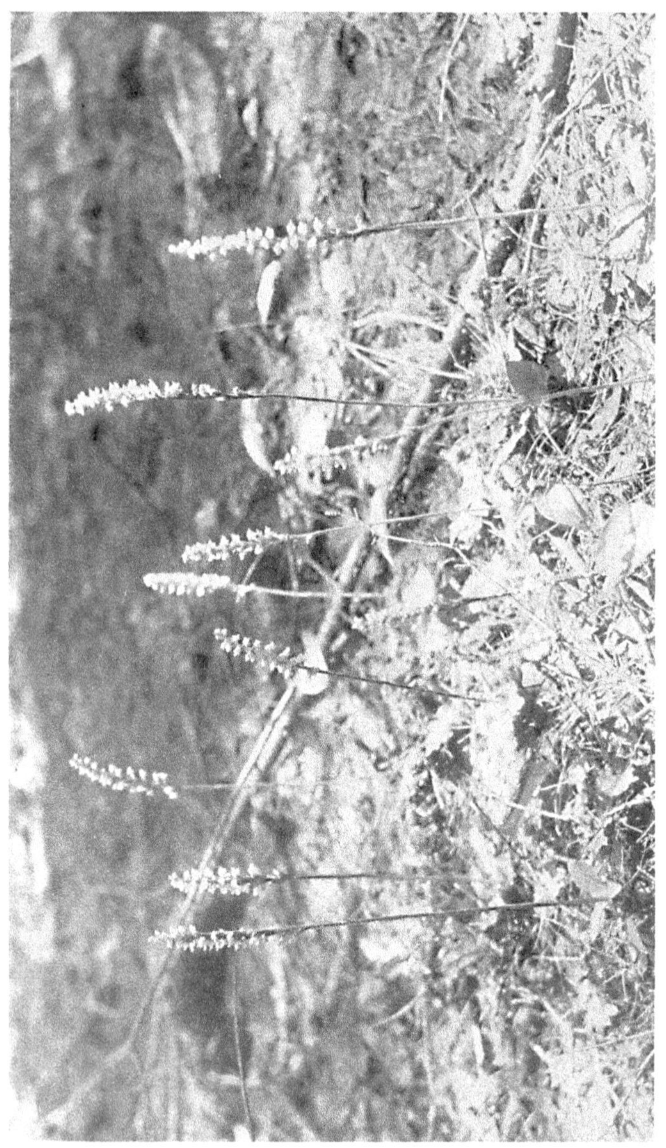

The Haunts of the Rattlesnake Plantain, amid the Pines and Spruces of the Domelet, Pownal, Vermont.

The generic name, *Listera*, is in honor of Martin Lister, 1638 (?)-1712, a correspondent of Ray.

Small orchids with fleshy-fibrous roots. Anther, 1. Labellum 2-cleft, longer than petals. Sepals and petals nearly alike. Flowers in terminal raceme, spurless. Anther erect, joined to column, without a lid. Pollinia, 2, united to gland, powdery. Stem 3-10 inches high. Leaves, 2, opposite, near the middle of the stem, 1-2 scales below.

Continental Range—In moist woodlands and boglands from Alaska, southward to Florida; westward to the Pacific coast. There are 12 species or more belonging to the north temperate zone which are closely related to species of *Gyrostachys* and *Peramium*, save in the herbaceous foliage.

North American species north of Mexico	8
New England species	3
Hoosac Valley species	1-2

New England species

1. *L. convallarioides* (Swartz) Torrey, 1800-1826.
2. *L. auriculata* Wiegand, 1899.
3. *L. cordata* (Linnaeus) R. Brown, 1753-1813.

1.—*Listera convallarioides* (Swartz) Torrey, 1800-1826†
Broad-Lipped Twayblade

The specific name, *convallarioides*, refers probably to a supposed resemblance of this species to *Convallaria*, the Lily-of-the-Valley.

Small woodland orchid, with fleshy-fibrous roots. June 9th-August 17th.

Flowers purplish-yellow, 3-15 spurless, subtended by acute bracts, 1/4-1/3 inch long. Labellum wedge-shaped, broader than sepals or petals, produced into 2 distinct lobes at the apex, notched in center, ornamented with tooth at the base. Sepals and petals linear-lanceolate. Stem 4-10 inches high. Leaves 2, nearly orbicular in the middle of stem, opposite, smooth, obtuse at the apex, 3-9 nerved.

Continental Range—From the wooded coast Unalaska, southward throughout the Canadian provinces, to North Carolina; westward to the fir-forests of Idaho, Washington, Wyoming, Sierra Nevada Mountains, and the Bay region of California.

† Species not reported for Hoosac Valley region, although native of Vermont.

New England Range—Maine, common; New Hampshire, frequent; Vermont, infrequent; Massachusetts, doubtfully reported.

2.—*Listera auriculata* Wiegand, 1899†
Auricled Twayblade

The specific name, *auriculata*, from *auriculum*, ear, refers to the auricled base of the labellum and leaves of this species.

Small cedar swamp orchid, with fibrous roots. July.

Flowers, many in slender raceme. Labellum slightly ciliate, oblong, broadest at the auricled base, cleft 1/4-1/3 its length. Sepals lanceolate; petals oblong-linear, longer than the ovary, spreading, obtuse. Stem 4-7 inches high, slender. Leaves large, oval, or elliptic-ovate, borne above the middle of the stem.

Continental Range—From Quebec, southward to Maine and New Hampshire.

New England Range—Maine, rare; New Hampshire, rare.

3.—*Listera cordata* (Linnaeus) R. Brown, 1753-1813.
Heart-Leaved Twayblade

The specific name, *cordata*, refers to the cordate or heart-shaped leaves.

Small, moist woodland orchid, with fibrous roots. June 27th-July 30th-August 8th.

Flowers minute, purplish, in a loose raceme, 1/2-2 inches long, 4-20 flowered, pedicels bracted, 1/6 inch long. Labellum 2-cleft, narrow, with a tooth on each side at the base. Sepals and petals oblong-linear. Stem slender, 3-10 inches high. Leaves 2, heart-shaped, or cordate 1/2-1 inch long.

Continental Range—From Alaska, southward to Sierra Nevada Mountains in the Pacific region; and New Jersey, and Pennsylvania, in the Atlantic region. It appears to be more of a northern plant than *L. convallarioides*.

New England Range—Maine, common; New Hampshire, common; Vermont, common; Massachusetts, infrequent; Rhode Island, rare; Connecticut, rare.

† Species not reported for Hoosac Valley region, although native of northern New England.

VIII

Peramium

Salisbury, 1812
(Goodyera R. Brown, 1813)
Rattlesnake Plantain

The origin of the generic name, *Peramium*, is not given in the original description. It may come from *Ammon, Amen*, the Egyptian Sun-God of Life, since to the blotched leaves of this genus the Indians attribute great powers, as a remedy against the deadly amniotic poison received from the rattlesnake's bite.

Orchids with erect bracted scapes, and fleshy-fibrous roots. Anther 1. Labellum concave, or sessile roundish-ovate. Lower sepals free, upper one united with petals into a galea. Flowers in densely-flowered, or 1-sided bracted spikes. Anther erect, attached to column, without a lid. Pollinia 2, 1 in each anther-sac, composed of angular grains attached to small disk, cohering with top of stigma. Stems or scapes bracted, 5-20 inches high. Leaves several, basal, blotched, with beautiful network of white, green or yellow, resembling a snake's skin. Seed-capsule erect, nearly always maturing. Seedlings abundant in *conifer* shades.

Continental Range—From Alaskan southward to Florida; westward to Minnesota and California. There are 25 or more species of this genus ranging in the temperate and tropical regions of the world.

North American species north of Mexico	5
New England species	4
Hoosac Valley species	3-4

New England species

1. *P. repens* (Linnaeus) Salisbury, 1753-1812.
2. *P. pubescens* (Willdenow) MacMillan, 1805-1892.
3. *P. Menziesii* (Lindley) Morong, 1840-1894.
4. *P. ophioides* (Fernald) Rydberg, 1899-1901.

1.—Peramium repens (Linnaeus) Salisbury, 1753-1812
Small One-Sided Goodyera—Net-Leaf Rattlesnake Plantain

The specific name, *repens*, refers to the creeping roots of this species.

Small *conifer*, woodland orchid, with fleshy-fibrous roots. July 19th-August 5th-30th.

Flowers white, on *1-sided* spike, 1/6-1/4 inch long. Labellum saccate, recurved, narrowed at apex; column short. Stem 5-10 inches high. Leaves ovate, basal in a rosette, pointed, yellowish-green, 1/2-1 1/4 inch long, 1/3-2/3 inch wide, blotched with white or lighter yellowish-green; several bracts above. This species intergrades with other New England species of *Peramium*, so as to make their designation difficult in many stations.

Continental Range—From Nova Scotia, possibly Alaska (?), southward to Florida; westward to Minnesota, South Dakota, and Colorado. Ascends 5000 feet altitude in Virginia.

New England Range—Maine, common; New Hampshire, common; Vermont, common; Massachusetts, infrequent; Rhode Island, very rare; Connecticut, rare.

2.—*Peramium pubescens* (Willdenow) MacMillan, 1805-1892
Downy Rattlesnake Plantain—Canker-Root

The specific name, *pubescens*, refers to the hirsute or downy leaves and scape of this species.

Small *conifer* woodland orchid, with fleshy-fibrous roots. June 15 Virginia (Curtiss); May 8th Wisconsin; July 4th-September, New England.

Flowers greenish-white, in densely-flowered (*not 1-sided*) spike. Labellum saccate, apex recurved, obtuse. Lateral sepals ovate; petals and upper sepal arching in an ovate galea. Stem 6-20 inches high, clothed with hairs, much more hirsute, or downy, than *P. repens*. Leaves basal, in a rosette, 1-2 inches long, strongly blotched with greenish-white, 5-10 scales above. The Creeping Goodyera (*P. repens*) intergrades with this species and causes confusion in designation. *P. pubescens* spikes are *not 1-sided*, the rosette of leaves are of a bluer velvety green, blotched with a purer white network, while *P. repens* rosette of leaves is dull yellowish-green.

Continental Range—From Ontario, Newfoundland, southward to Florida; westward to Minnesota. Ascends 4000 feet altitude in North Carolina.

New England Range—Maine, common; New Hampshire, common; Vermont, common; Massachusetts, frequent; Rhode Island, rare; Connecticut, common.

3.—*Peramium Menziesii* (Lindley) Morong, 1840-1894†
Menzies' Rattlesnake Plantain

† Species not reported for Hoosac Valley region, although native of northern New England.

The specific name, *Menziesii*, refers to the dedication of this species, in honor of the explorer and botanist Menzies.

Small *conifer* woodland orchid, with fibrous-fleshy roots. June-July 21st-August 20th-September 15th.

Flowers greenish-white, spike *not 1-sided*. Labellum scarcely saccate, swollen at base, apex narrower, and recurved. Anther ovate, pointed; the buds, flowers, and leaves are all conspicuously acute. Stem 8-15 inches high. Leaves basal in rosette, bracts above, often without blotches of white; resembling *P. pubescens* very much, although the basal leaves are stiffer and acute at apex and base. The *Peramiums* intergrade with each other, confusing their specific characters.

Continental Range—From Loring, Chilcat, southeast coast of Alaska, southward to Lake Huron, Quebec, Maine, Vermont (?), and New York; westward to Arizona and California. Ascends 9500 feet altitude In Arizona.

New England Range—Maine, rare; New Hampshire, rare; Vermont, doubtfully reported; Massachusetts, doubtfully reported.

4.—*Peramium ophioides* (Fernald) Rydberg, 1899-1901
White-Blotched Rattlesnake Plantain

The specific name, *ophioides*, comes from *ophis*, a serpent, and *oides*, like, referring to the blotched leaves resembling a snake's skin.

Small cold mossy woodland orchid with thick fleshy-fibrous roots. July-September.

Flowers greenish-white; galea concave with a short, strongly recurved tip. Labellum deeply saccate, with recurved margins and tip. Anther blunt. Scape 4-8 inches high, glandular-pubescent. Leaves basal in rosette, several; leaf-blade broadly ovate, dark green, usually with the white blotches most conspicuous along the cross-veins. Spike of flowers 1-sided, loosely arranged. A variety of *P. repens*, with which it is confused.

Continental Range—From Prince Edward Island to Manitoba, southward to North Carolina, in company with *P. repens* and *P. pubescens*.

New England Range—Vermont, frequent; Massachusetts, common.

IX

ACHROANTHES

Rafinesque, 1808
(*Microstylis* Nuttall, 1818)
Adder's-Mouth

The generic name, *Achroanthes*, refers to the green flowers of this genus.

Small orchids with solid bulbs. Anther, 1. Labellum cordate, or eared, at the base, embracing the column. Sepals free; petals linear, spreading. Flowers minute white or greenish, in a terminal raceme. Anther erect between the auricles, 2-celled. Pollinia, 4, 2 in each anther-sac, smooth and waxy, without caudicles or glands, and cohering at summit. Stem, 4-10 inches high. Leaf 1, with several scales at base of stem. Seed-capsule oval, or globose.

Continental Range—From Alaska, southward to Florida; westward to Nebraska and Arizona, where a species seeks 9500 feet elevation on Mount Graham. There are 70-80 species reported in the temperate and tropical regions of the world.

North American species north of Mexico 7
New England species 2
Hoosac Valley species 2

New England species

1. *A. monophylla* (Linnaeus) Green, 1753-1891.
2. *A. unifolia* (Michaux) Rafinesque, 1803-1808.

1.—*Achroanthes monophylla* (Linnaeus) Green, 1753-1891
White Adder's-Mouth

The specific name, *monophylla*, refers, inappropriately, to the one leaf, since each of our New England species are 1-leaved.

Small woodland orchid, with bulbous root. June 20th-July 25th-August 2d.

Flowers whitish, in a club-like raceme, 1-3 inches long, 1/4-1/2 inch thick; flowers 1/12 inch long, minute, pedicels nearly erect, subtended by bracts 1/6 inch long. Labellum ovate, acuminate, notched on sides. Sepals and petals acute, narrow. Stem, 4-6 inches high. Leaf, 1, sheathing at its base, 1-2 inches long, distinguished from following species, *A. unifolia*, by being near base of stem instead of middle.

Continental Range—From New Brunswick, Nova Scotia, southward to Vermont, doubtfully reported for New Bedford,

The Green Adder's-Mouth. (*Acroanthes unifolia.*)

Massachusetts; westward to Oneida, New York, Pennsylvania, Minnesota, Indiana, and Texas.

New England Range—Maine, infrequent; New Hampshire, rare; Vermont, Pownal, Swamp of Oracles (Marcus White), rare; Massachusetts, Berlin, and Spencer (Dr. G. E. Stone); Mount Greylock bog, North Adams (Marcus White), rare.

2.—*Achroanthes unifolia* (Michaux) Rafinesque, 1803-1808
Green Adder's-Mouth

The specific name, *unifolia*, one-leaved, refers, like the preceding species, to the 1 leaf.

Small damp woodland orchid, with bulbous root. May-June 26th (South)-July 25th-August 10th-September (North).

Flowers in club-shaped raceme 1-3 inches long, 1 inch thick; greenish, minute, 1/12 inch long, pedicels slender, spreading, 1/4-1/2 inch long. Labellum broad, 3-toothed at apex. Sepals oblong. Stem, 4-10 inches high. Leaf, 1, near the middle of stem, clasping, nearly orbicular. Seed-capsule oval, drooping.

Continental Range—From Newfoundland, Ontario, southward to Florida; westward to Missouri and Minnesota. Ascends 4000 feet altitude in North Carolina.

New England Range—Maine, frequent; New Hampshire, frequent; Vermont, Pownal (Grace G. Niles), rare; Massachusetts, Mount Greylock pasture, North Adams (Marcus White), rare; Rhode Island, rare; Connecticut, rare.

X.
LEPTORCHIS

Thouars, 1808
(*Liparis* Richard, 1818)
Lily-Leaved Twayblade

The generic name, *Leptorchis*, comes from the Greek, referring to a slender orchid. The former name, *Liparis*, referred to the smooth and shining leaves of these plants.

Small low orchids with bulbous roots. Anther, 1. Labellum nearly flat, often bearing 2 tubercles above the base. Sepals and petals spreading, petals usually narrow. Flowers in a terminal raceme, numerous and showy. Pollinia, 4, 2 in each anther-sac, smooth and waxy, slightly united, without stalks or glands. Stem or scape, 2-10 inches high. Leaves, 2, basal, broad, oval and shining, with several sheathing scales at base. Two or three seasons bulbs adhere to the latest bulb, and in time wither away. Seed-capsule long, erect, club-shaped.

The Large Twayblade. (*Leptorchis liliifolia.*)

Continental Range—In moist woodlands and along springy roadside banks. From Canada, New England southward to Georgia; westward to Iowa, Minnesota, and Washington. There are about 100 species of this genus, distributed in the temperate and tropical regions of the world.

North American species north of Mexico	2
New England species	2
Hoosac Valley species	2

New England species

1. *L. liliifolia* (Linnaeus) Kuntze, 1753-1891.
2. *L. Loeselii* (Linnaeus) MacMillan, 1753-1892.

1.—*Leptorchis liliifolia* (Linnaeus) Kuntze, 1753-1891
Large Lily-Leaved Twayblade

The specific name, *liliifolia*, refers to the lily-leaves of this species.

Small, moist woodland orchid, with bulbous onion-like roots, more or less exposed. May 17th-June 1st -July 16th.

Flowers purplish-green in loose terminal raceme, many-flowered, and showy. Labellum purple, wedge-obovate. Sepals and petals narrow and reflexed. Stem or scape, 4-10 inches high. Leaves, 2, basal, oval, 2-5 inches long, smooth shining emerald-green. Seed-capsule club-shaped, erect.

Continental Range—From Maine southward to Georgia and Alabama; westward to Minnesota.

New England Range—Maine, rare; New Hampshire, infrequent; Vermont, infrequent; Massachusetts, infrequent; Rhode Island, rare; Connecticut, frequent.

2.—*Leptorchis loeselii* (Linnaeus) MacMillan, 1753-1892
Loesel's Green Twayblade—Fen-Orchis

The specific name, *Loeselii*, refers to the dedication of this species in honor of the botanist Loesel.

Small damp thicket or dry sandy broadside orchid, with onion-like bulbous roots; old bulbs adhering to latest bulb, somewhat exposed. May 30th-June 25th-July 27th.

Flowers greenish, minute 1/6-1/4 inch long, in a few-flowered raceme, smaller than preceding species, *L. liliifolia*; one of the smallest native orchids. Labellum pointed, obovate, apex incurved. Sepals narrow, lanceolate. Petals reflexed, linear. Stem or scape 2-8 inches high, 5-7 ribbed. Leaves 2, basal, 2-6 inches long, stiff; strongly veined, resembling Plantain leaves. Seed-capsules erect, wide-angled.

Continental Range—From Nova Scotia southward to Maryland, Kentucky; westward, to Minnesota and Washington.

New England Range—Maine, common; New Hampshire, frequent; Vermont, common; Massachusetts, infrequent; Rhode Island, not reported; Connecticut, rare.

XI

CALYPSO†

Salisbury, 1807

Beautiful Calypso—Northern Calypso

The generic name, *Calypso*, refers to the dedication of this genus to the goddess Calypso. Its Greek signification is not only, as Salisbury wrote, to "cover and conceal" the stigma of the species, but also to preserve a poetical analogy between this botanical goddess, so difficult of access, and the secluded goddess of Silence, whose Isle of Ogygia was fabled to be miraculously protected from observation by navigators.*

Small bogland orchid with solid bulbous and coralloid roots. Anther 1. Labellum shoe-shaped, saccate, 2-parted at the apex. Sepals and petals free, similar in texture. Flowers 1, large, terminal pendulous, bracted, resembling a Lady's Slipper (*Cypripedium*). Linnaeus wrongly designated this species *Cypripedium bulbosum* in 1753. Stem or scape 3-6 inches high. Leaf 1, hyemal, appearing as an autumnal leaf about September 2d,‡ sheathed above by 2-3 scales. Anther lid-like below the summit of column. Pollinia 2, 2-parted, without caudicles, waxy, sessile, on a thick gland. Seed-capsule about 1/2 inch long, many-nerved.

Continental Range—From Alaska, Labrador, southward, to Middlebury, Vermont, and doubtfully reported for Pelham, Massachusetts; westward to California and New Mexico. First collected in the United States in the State of Vermont, at Charleston and Morgan by the botanist Carey, who resided at Bellows Falls in 1831-1833. A monotypic species ranging in cooler portions of north temperate zone, in Europe, Asia, and North America, assuming slight varietal changes in different regions.

---

† Genus not reported for Hoosac Valley region, although native of Vermont.

* Salisbury, *Pard. Lond.*, pl. 89. 1807.

‡ Henry Baldwin, *Orchids of New England*, 93. 1894.

From lithograph in Meehan's *Native Flowers and Ferns of the United States*, 1: 1878. By permission.

Northern Calypso. (*Calypso bulbosa.*)

Calypso, goddess of an ancient time,
(I learn it not from any Grecian rhyme.
And yet the story I can vouch is true.)
Beneath a pine tree lost her dainty shoe.
. . . . .
The goddess surely must have been in haste,
Like Daphne fleeing when Apollo chased,
And leaving here her slipper by the way,
Intends to find it on another day.

W. W. Bailey

North American species 1
New England species 1
Hoosac Valley species 0

New England species

1. *C. bulbosa* (Linnaeus) Oakes, 1753-1842.

1.—Calypso *bulbosa* (Linnaeus) Oakes, 1753-1842†
Beautiful Calypso—Northern Calypso

The specific name, *bulbosa*, refers to the bulbous root of this species, which was originally confounded by Linnaeus in 1753 as a bulbous *Cypripedium*, and later placed under its generic designation *Calypso* by Salisbury in 1807.

Small sphagnous bogland or *conifer* woodland orchid, with bulbous and coralloid roots. April 19th-May 3d-June 15th-July 12th.

Flowers terminal, variegated with purple-pink, yellow, or white; shoe-shaped or saccate, resembling a Lady's Slipper, with which genus it was confused by Linnaeus in 1753. Labellum large, saccate or shoe-shaped pink-purple, 2-parted at apex, with patch of yellow (or white) woolly hairs near the point of division, spreading. Sepals and petals free, similar in texture. Stem or scape 3-6 inches high. Leaf 1, autumnal, appearing about September 2d, hyemal, basal, sheathed above by 2-3 scales. Seed-capsule 1/2 inch long, many-nerved.

Continental Range—From Sitka, Alaska, Labrador, southward to Middlebury, Vermont, and possibly as far south as Pelham, Massachusetts; westward to Humboldt Bay, mouth of Russian River, California, and northeastern New Mexico.

First collected in the United States in Vermont by the botanist Carey in 1831-1833. The Rocky Mountain *Calypso* appears to be distinguished from the eastern *Calypso* by producing a beard of white instead of yellow hairs at the point of division of the labellum. This varietal form is designated *Calypso occidentalis* (Holzinger) Heller.

Robert Brown, as early as 1813, attempted to establish a distinct species in the American *Calypso*, from that of the European and Asiatic forms. He designated the North American form, *Calypso Americana*. Neither Lindley nor Hooker approved of this distinction, Hooker remarking that the species even in the same country may vary in structure or

† Species not reported for Hoosac Valley region, although native of Vermont.

colors, but not permanently enough to designate it specifically. Smith, Richard and Lindley later agreed with Dr. Hooker in "considering the American, European and Asiatic *Calypso* the same."

Calypso bulbosa is the only species of this genus in the north temperate zone, and is nearly related to the section of *Pleiones* of genus *Cœlogyne*, meaning "two-lipped" or 2-parted at the apex of the labellum. *Cœlogyne* is a native of Asia, and many of the closely allied *Pleiones* are alpine-orchids, their large rose-colored or cream-colored flowers clinging to the branches of sturdy oaks at an altitude of 7500 feet in latitude 30° North. *Calypso* also seeks the colder lands, of the *conifer* forests of Alaska and Labrador, in latitude 54°-69° North; while in the Rocky Mountain region it is found at an elevation of 4000-5000 feet above sea level.

New England Range—Maine, frequent; New Hampshire, infrequent; Vermont, frequent northward; Massachusetts, doubtfully reported.

XII
CORALLORHIZA

R. Brown, 1813
Coral-Root

The generic name, *Corallorhiza*, refers to the coral-like masses of the roots of this genus.

Scapose orchids, saprophytes or root-parasites, with large masses of coralloid branching roots. Anther 1. Labellum 1-3 ridged. Sepals and petals equal; 1-3 nerved, lateral sepals united at the base with the foot of the column, forming a short spur. Flowers several in a terminal raceme, purplish, yellowish and white. Anther terminal, operculate. Pollinia 4, in 2-pairs, free, soft and waxy. Stem or scape 4-20 inches high. Leaves all reduced to scales. Seed-capsule oblong, drooping when ripe.

Continental Range—In rich woodlands. From Alaska, southward to Florida; westward to the Pacific region. There are about 15 species of this strange genus in the north temperate zone of the world. These species are destitute of green or any distinct form of foliage; their roots are without form, assuming coral-like masses, which draw nourishment for the plants from humus—the rich decay of dead roots and trees in the soil. These plants are known as saprophytes or root parasites.

North American species north of Mexico 9
New England species 4
Hoosac Valley species 3

The Coral-Root. (*Corallorhiza.*)

New England species

1. *C. corallorhiza* (Linnaeus) Karst, 1753-1880-1883.
2. *C. odontorhiza* (Willdenow) Nuttall, 1805-1818.
3. *C. Wisteriana* Conrad, 1829.
4. *C. multiflora* Nuttall, 1823.

1.—*Corallorhiza corallorhiza* (Linnaeus) Karst, 1753-1880-1883
Early Coral-Root

The specific name, *corallorhiza*, refers to the coral-like roots of the species and genus.

Scapose rich woodland orchid, with coral-like roots. May 11th-June 28th-July 12th-August 5th (North)-September-October (South). A vernal orchid, North, and an autumnal species in Georgia. The dates of flowering for the *Corallorhizas* are variable and not standard.

Flowers, 3-12 in a raceme 1-3 inches long, dull purple, about 1/2 inch long on short, minutely bracted pedicels. Labellum whitish, oblong, 2-toothed, shorter than petals, spur a small protuberance adnate to the summit of ovary. Sepals and petals narrow. Stem or scape, 4-12 inches high. Leaves reduced to 2-5 sheathing scales. Seed-capsule drooping.

Continental Range—From Greenland, and Kotzebue Sound, Unalaska, southward throughout Canada to Georgia; westward to Washington and Oregon. Ascends 7600 feet in Yellowstone Park, where it is rather common.

New England Range—Maine, common; New Hampshire frequent; Vermont, common; Massachusetts, common; Rhode Island, not reported; Connecticut, infrequent.

2.—*Corallorhiza odontorhiza* (Willdenow) Nuttall, 1805-1818
Small-Flowered Coral-Root—Dragon-Claw—Late Coral-Root—Crawley-Root

The specific name, *odontorhiza*, comes from the Greek, *odons*, a tooth, and *hiza*, a root, referring to the tooth-like shape of the coralloid roots.

Small slender woodland orchid, with coral-like masses of roots. February-March-May (South); July-August-September 6th-October 15th (North).

Flowers, 6-20, purplish, in raceme 2-4 inches long. Labellum oval, denticulate, narrowed at base; not notched, whitish; spur small, adnate to top of ovary. Sepals and petals lanceolate, marked with purple lines. Stem sheathed with 3-4 scales, 6-15 inches high. Confused with *C. corallorhiza* northward.

Continental Range—From Ontario, Canada; Halifax, Nova Scotia, southward to Alabama, Florida, and Texas; westward to Illinois and Indiana. Ascends 3000 feet in North Carolina.

New England Range—Maine, rare; New Hampshire, rare; Vermont, rare; Massachusetts, Cole's Grove, Williamstown (Cheney), rare; Rhode Island, rare; Connecticut, frequent.

3.—*Corallorhiza wisteriana* Conrad, 1829†
Wister's Coral-Root

The specific name, *Wisteriana*, refers to the dedication of this species in honor of the botanist Wister.

Slender woodland orchid, with coral branching roots. March 1st, Florida (Curtiss); Alabama, February-May.

Flowers whitish, 6-15, in spiked raceme 2-5 inches long; pedicels, erect and slender. Labellum broad and oval, white, clawed, with spots of crimson; notched at apex, differing in this from *C. odontorhiza*, which is not notched at the apex, but projects acutely; spur a conspicuous protuberance adnate to top of the ovary. Stem 8-16 inches high. Leaves reduced to several sheathing scales. Seed-capsule oblong drooping, when ripe.

Continental Range—From Massachusetts southward to Florida and Texas; westward to Ohio, taking much the same range, and flowering at the same time as *C. odontorhiza*.

New England Range—Massachusetts, rare.

4.—*Corallorhiza multiflora* Nuttall, 1823
Spotted Large Coral-Root

The specific name, *multiflora*, refers to the multiplying of both flowers and plants in many stations.

Tall woodland orchid, with large masses of coralloid roots. May (Canada)-June 20th-July (Maine); August 14th (Massachusetts); September 15th (Connecticut).

Flowers, 10-30, brownish-purple, in spiked raceme 2-8 inches long; pedicels short; flowers 1/2-3/4 inch long. Labellum white, spotted with purple, oval deeply 3-lobed, central lobe broad; side lobes narrow, apex curved. Spur manifest. Sepals and petals linear-lanceolate. Stem 2-20 inches high, purplish. Leaves reduced to several appressed scales. Seed-capsule oblong, drooping when ripe.

---

† Species not reported for Hoosac Valley region, although native of Massachusetts.

Continental Range—From Newfoundland, Nova Scotia, southward to Florida; westward to California. Ascends 2500 feet altitude in Montana (Tweedy).

New England Range—Maine, rare; New Hampshire, rare; Vermont, rare; Massachusetts, frequent; Rhode Island, rare; Connecticut, frequent.

XIII

TIPULARIA†

Nuttall, 1818
Crane-Fly Orchis

The generic name, *Tipularia*, refers to the flowers resembling insects of genus *Tipula*.

Slender scapose orchids, with solid bulbous roots; several bulbs, or generations connected by offsets. Anther 1. Labellum 3-lobed produced into a long spur backwardly. Sepals and petals similar, spreading. Flowers in a long, loose terminal raceme. Anther terminal, operculate, 2-celled. Pollinia, 4, 2 in each anther-sac, ovoid, waxy, separate, affixed to short stipe, glandular at base. Stem 15-20 inches high. Leaf 1, basal, arising in autumn, about September 14th, hyemal after the flowering-scape has perished. Seed-capsule 1/2 inch long, 6-ribbed.

Continental Range—From Brattleboro and Bellows Falls, Vermont, to New Jersey, Alabama, and Florida; westward to Ohio, Michigan, and Indiana. Rare in sandy woods. There are but two known species reported for the world, the following, and another, native of Asia, in the Himalayan region. The American species is slightly distinguished from the Asiatic form by the blunt tip of its labellum.

North American species north of Mexico	1
New England species	1
Hoosac Valley species	0

New England species:

1. *T. unifolia* (Muhlenberg) B. S. P., 1813-1888.

1.—*Tipularia unifolia* (Muhlenberg) B. S. P., 1813-1888‡
(*Tipularia discolor*, Nuttall, 1818)
Crane-Fly Orchis

† Genus not reported for Hoosac Valley region, although native of Vermont.

‡ Species not reported for Hoosac Valley region, although native of Vermont.

The specific name, *unifolia*, refers to the 1 leaf produced by this orchid.

Alert, small moist sandy woodland or rocky hillside orchid, with solid irregular bulb or corm-like roots. Late July-August-October.

Flowers green, tinged with purple, 1/3-1/2 inch long, in a loose raceme, 5-10 inches long; pedicels filiform, bractless. Labellum 3-lobed, central lobe narrow, prolonged, dilated at apex, side-lobes short and triangular. Spur straight, slender, twice as long as sepals and petals, giving an insectean poise to the dull flowers. Sepals and petals narrow. Stem 15-20 inches high. Leaf 1, basal, reddish-purple, strongly veined; arising from solid bulb, about September 14th, hyemal. Seed-capsule 1/2 inch long, 6-ribbed.

Continental Range—From southern Vermont, New Hampshire, southward to Florida and Alabama; westward to Michigan and Indiana. More abundant South. Nowhere common.

New England Range—Vermont, rare; Massachusetts, rare.

XIV
LIMODORUM

Linnaeus, 1753
(*Calopogon* R. Brown, 1813)
Grass-Pink—Meadow-Gift

The generic name, *Limodorum*, comes from the Greek, signifying a meadow-gift.

Scapose orchids with solid bulbous roots. Anther 1. Labellum, hinged, arching above, and spreading; raised on a narrow stalk, dilated at the apex, bearded on the upper side with long club-shaped hairs. Sepals and petals nearly alike, separate and spreading. Flowers fragrant, pink-purple, 3-15 in a loose terminal raceme, seed-capsule (ovary) straight. Anther terminal, operculate, and sessile. Pollinia 2, 1 in each anther-sac, loosely granular. Stem or scape straight, not twisting as usual in other orchids, 1-1 1/2 foot high. Leaf 1, grass-like blade, appearing first season, and followed next year by scape of flowers. Seed-capsule erect, oblong, and straight.

Continental Range—From Newfoundland, Canada, southward to Florida; westward to Minnesota and Arkansas. There are 4 species of this beautiful genus endemic only to the Atlantic region.

A peculiar character of this genus lies in the ovary and stem being straight, causing thereby the labellum to arch above instead of drooping below the organs of fertilization, as instanced in *Orchis* and *Cypripedium*. Seedlings appear numerous in many swamps.

The Grass-Pink. (*Limodorum tuberosum.*)

A beautiful grassy-leaved orchid found in company with the dainty Rose Pogonia, and frequently with the rarer Arethusa in wild cranberry marshes.

North American species north of Mexico 4
New England species 1
Hoosac Valley species 1

New England species

1. *L. tuberosum* Linnaeus, 1753.
 (*Calopogon pulchellus* R. Brown, 1813)

1.—*Limodorum tuburosum* Linnaeus, 1753
(*Calopogon pulchellus* R. Brown, 1813)
Grass-Pink—Meadow-Gift

The specific name, *tuburosum*, refers to the tuberous or bulbous roots of this orchid.

Beautiful grassy sphagnous meadow orchid, with bulbous roots. May 15th-June 1st-July 20th-August 1st.

Flowers, 3-15 pink-purple, 1 inch long, subtended by acute bracts in spiked raceme 4-15 inches long. Labellum hinged, arching above, owing to ovary and stem being straight; broad, triangular at apex; bearded on the upper side with yellow, orange, and rose-colored club-shaped hairs. Sepals and petals acute, ovate-lanceolate, similar in texture and color. Stem or scape 1-1 1/2 feet high. Leaf 1, linear-lanceolate, 8-12 inches long, grass-like with several scales below. Seed-capsule straight, erect, oblong.

Continental Range—From Newfoundland, Nova Scotia to the shores of Lake Huron, southward throughout New England to Florida; westward to western Texas and Minnesota.

New England Range—Maine, common; New Hampshire, common; Vermont, common; Massachusetts, common; Rhode Island, common; Connecticut, common.

XV
APLECTRUM

Nuttall, 1818†
Putty-Root—Adam-and-Eve

The generic name, *Aplectrum*, comes from the Greek meaning without a spur.

Scapose orchids with bulb or corm-like roots. Anther 1. Labellum 3-lobed, shorter than petals. Sepals and petals about 1/2 inch long.

† Genera not reported for Hoosac Valley region, although native of Vermont.

Flowers without a spur, dull yellowish-brown in a loose raceme. Anther borne a little below the summit of column. Pollinia 4, lens-shaped. Stem or scape 1-2 feet high. Leaf 1, basal, arising from side of scape; several corms adhering to latest bulb; leaf develops in late autumn, about September 9th, hyemal; several sheathing scales above. Seed-capsule oblong, ovoid, angled.

Continental Range—From Ontario, southward to Georgia and Alabama; westward to Minnesota, Washington, Oregon, Idaho, and probably in northern California. Rather rare and local.

North American species north of Mexico...........2
New England species..............................1
Hoosac Valley species............................0

New England species

1. *A. spicatum* (Walter) B. S. P., 1788-1888.

1.—*Aplectrum spicatum* (Walter) B. S. P., 1788-1888†
(*Aplectrum hyemale* Nuttall, 1818)
Putty-Root—Adam-and-Eve‡

The specific name, *spicatum*, refers to the flowers growing in a *spica*, or spike.

Tall, spiked damp sandy woodland or bogland orchid, with bulbous or corm-like roots. May 22d-July 1st (Northern States); April 20th-July 1st (Southern States).

Flowers, 1-9 dull yellowish-brown, mixed with purple, 1 inch long, short-pediceled, in a loose raceme 2-4 inches long. Labellum shorter than petals, 3-lobed. Sepals and petals 1/2 inch long, linear-lanceolate. Stem or scape 1-2 feet high, producing 3 scales above the leaf. Leaf 1, basal, arising at side of

---

† Species not reported for Hoosac Valley region, although native of Vermont.

‡ The common name, Putty-Root, arose from the putty-like consistency of the adhesive substance of the old corms or bulbs; used to mend broken china. The name Adam-and-Eve originated with the colored folk in Georgia and Alabama. The bulbs are not disagreeable to eat if baked, and many confess to be fond of them. The colored people in the South are said to wear these bulbs as amulets, and attribute great favor to them in casting lots. By separating the offsets, they designate them Adam-and-Eve, as the rule may be, and placing them in a bowl of water decide their good or ill fortune in obtaining work, or a lover, according as Adam or Eve "pops up."

Epiphytes, or Air Plants.

A Corner in the Orchid House of the Botanical Gardens of New York City.

scape, from the latest bulb or corm; elliptic, 4-6 inches long, appearing about September 9th, hyemal—lasting through the winter.

Continental Range—From Ontario, southward to Georgia and Alabama; westward to Minnesota, Oregon, Idaho, Washington, and probably California.

This species, like those of *Corallorhiza*, is not a definite dated flowering orchid; in the Virginian ravines it blooms as early as April 20th while in Wisconsin, and Missouri it blooms as late as July 1st. The average date for New England is from May 22d-June 25th.

New England Range—Maine, rare; New Hampshire, rare; Vermont, rare; Massachusetts, rare; Rhode Island, not reported; Connecticut, rare.

The Pearly Mussels of Pennsylvania

Al Spoo
ISBN 1-930585-50-0
$59.95

All 66 species of freshwater mussels (extant and extirpated) in Pennsylvania are included in this full-color guide. A must for conservationists and wildlife enthusiasts in the Mid-Atlantic region.

A Year at the Shore

Philip Henry Gosse
ISBN 1-930585-51-9
$24.95

Gosse's classic work on British coastal marine life is back in print, with all 36 full-color plates.

COACHWHIP PUBLICATIONS

COACHWHIPBOOKS.COM

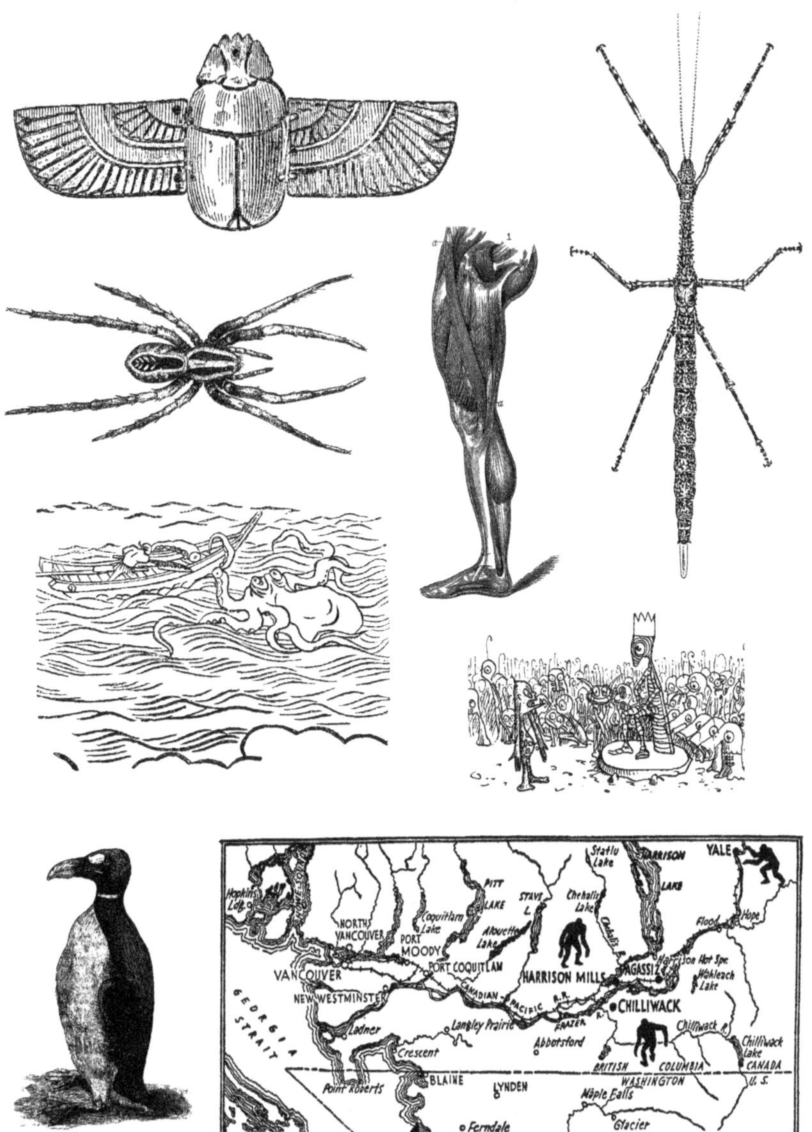

www.ingramcontent.com/pod-product-compliance
Lightning Source LLC
Chambersburg PA
CBHW041157280326
41927CB00019BA/3376